Ethical Dilemmas in Prenatal Diagnosis

Tamara Fischmann • Elisabeth Hildt
Editors

Ethical Dilemmas in Prenatal Diagnosis

Editors
Tamara Fischmann
Sigmund-Freud-Institut
Myliusstraße 20
60323 Frankfurt
Germany
dr.fischmann@sigmund-freud-institut.de

Elisabeth Hildt
Department of Philosophy
University of Mainz
Jakob-Welder-Weg 18
55099 Mainz
Germany
hildt@uni-mainz.de

ISBN 978-94-007-1395-6   e-ISBN 978-94-007-1396-3
DOI 10.1007/978-94-007-1396-3
Springer Dordrecht Heidelberg London New York

Library of Congress Control Number: 2011929871

© Springer Science+Business Media B.V. 2011
No part of this work may be reproduced, stored in a retrieval system, or transmitted in any form or by any means, electronic, mechanical, photocopying, microfilming, recording or otherwise, without written permission from the Publisher, with the exception of any material supplied specifically for the purpose of being entered and executed on a computer system, for exclusive use by the purchaser of the work.

Printed on acid-free paper

Springer is part of Springer Science+Business Media (www.springer.com)

Preface

The book "Ethical dilemmas in prenatal diagnosis" is based on the contributions to the international conference "Ambivalence of the technological progress in medicine" which was held in September 2008 in Frankfurt/Main, Germany. The focus of this conference was on technological developments in the life sciences which – often tacitly – confront us with new facets of a Faustian seduction. Are we "playing God more and more, as some contemporary critical authors of modernity are claiming? Achievements in genetic research produce ethical dilemmas which need to be the subject of reflection and debate in modern societies. Denial of ambivalences that ethical dilemmas arouse constitutes a threat to societies as well as to individuals. The European study "Ethical Dilemmas Due to Prenatal and Genetic Diagnostics" (EDIG, 2005–2008) investigated some of these dilemmas in detail in a field which is particularly challenging: prenatal diagnosis. When results from prenatal diagnosis show fetal abnormalities, women and their partners are confronted with ethical dilemmas regarding: the right to know and the right not to know; decision-making about the remainder of the pregnancy and the desire for a healthy child; responsibility for the unborn child, for its well-being and possible suffering; life and death.

With this book we would like to contribute to this ongoing and demanding discussion from an ethical, psychoanalytical and medical perspective. The book presents a compilation of some of the results of the EDIG study providing an insight into the interdisciplinary nature of the research. In a sense, this book can be considered to be a continuation of the volume *The Janus Face of Prenatal Diagnostics. A European Study Bridging Ethics, Psychoanalysis and Medicine* edited by Marianne Leuzinger-Bohleber, Eve-Marie Engels, and John Tsiantis and published by Karnac Books in 2008.

Marianne Leuzinger-Bohleber, the coordinator of the EDIG study, introduces the subject-matter by giving a short overview of the research project. In particular, she highlights clinical-psychoanalytical findings and discusses protective as well as risk factors that may help medical staff identify couples who particularly need support and help during the process of having to decide on the life or death of their unborn child. Then Nicole Pfenning-Meerkötter discusses the management of such a large and complex undertaking as the EDIG research project and the challenges posed to the scientists involved.

In the following chapter, Tamara Fischmann presents empirical results of the EDIG study, focusing on distress resulting from prenatal diagnosis and related ethical issues. She describes the study design and population, as well as the impact on women of the testing process. Katrin Luise Läzer scrutinizes the subjective attitudes of women towards prenatal testing by analyzing empirical EDIG data with a qualitative approach. She explores women's motives for undergoing prenatal testing and their feelings towards the process. A positive test result is a disturbing experience for women and their partners who find themselves at the limit of their emotional capabilities. Elisabeth Hildt focuses on this difficult issue from an ethical point of view studying the responses given by women to the open questions about their decision making processes.

After reviewing developments in prenatal testing, the chapter by Helen Statham and Joanie Dimavicius provides important information which may help professionals give optimum support and care to those undergoing prenatal diagnosis. In particular, a number of useful Internet-based resources for professionals and couples are presented. Astrid Riehl-Emde and colleagues summarize some 10 years of research in Germany, which has focused on collaboration between different professional groups involved with women during prenatal diagnosis. The authors suggest that although there have been definite advances, there is still a need for better coordination of information and care for pregnant women and their partners.

Anders Nordgren reflects, in his chapter, on policy frameworks for prenatal genetic counselling and the implications that empirical findings, such as those from the EDIG study, may have for informing decisions about policy. Another facet of prenatal diagnosis and genetic counselling, highlighted here by László Kovács, is the balancing of risk versus the need for certainty. His chapter analyzes differences in the practical and ethical concerns of health care professionals and of pregnant women and he suggests that ethically sound solutions have to integrate both perspectives. Dierk Starnitzke's contribution draws on his experience as speaker of the board of one of the largest institutions for disabled people in Germany. He discusses counselling with respect to termination or continuation of a pregnancy after a positive test result. Yair Tzivoni argues that a consideration of the unconscious in the decision-making process could promote a more ethical and responsible way of working through this process both for the person who has to decide and for the doctor/counsellor.

Regina Sommer then draws on the Aristotelian model of decision making in order to explain limitations in the choice a pregnant woman will make when confronted with an abnormal prenatal test result. From an applied ethics perspective, Göran Collste considers moral aspects of decision-making by drawing on four interviews with women who had been confronted with difficult choices through the information provided from prenatal diagnosis. Autonomous decision-making is also scrutinized in a clinical case vignette from Maria Samakouri and colleagues. This emphasizes the ways in which certain personality traits and past life experiences may influence decisions concerning prenatal testing. And finally, Stephan Hau poses the question 'Different cultures – different ethical dilemmas?' He highlights an ongoing public debate on ethical questions in the field of prenatal diagnosis in Sweden.

This collected volume would not have been possible without the research, effort, enthusiasm and patience of the partners of the EDIG study. We thank all those who participated in the conference and who contributed a chapter to this book, as well as the additional authors. We are particularly grateful to Helen Statham and Joanie Dimavicius for their elaborated and thoughtful English editing of the manuscripts and for stimulating discussions.

And last but not least we would like to thank all the women and their partners who so willingly agreed to participate in this study and gave us so much insight by talking to us and by completing all those questionnaires.

January 2011

Tamara Fischmann
Elisabeth Hildt

Contents

Preface v

Contributors xi

1 **Ethical Dilemmas Due to Prenatal and Genetic Diagnostics. An Interdisciplinary, European Study (EDIG, 2005–2008)** 1
Marianne Leuzinger-Bohleber

2 **Managing Complex Psychoanalytic Research Projects Applying Mapping Techniques – Using the Example of the EDIG Study** 35
Nicole Pfenning-Meerkötter

3 **Distress and Ethical Dilemmas Due to Prenatal and Genetic Diagnostics – Some Empirical Results** 51
Tamara Fischmann

4 **Reconstruction of Pregnant Women's Subjective Attitudes Towards Prenatal Diagnostics – A Qualitative Analysis of Open Questions** 65
Katrin Luise Läzer

5 **Prenatal Testing: Women's Experiences in Case of a Conspicuous Test Result** 75
Elisabeth Hildt

6 **Caring for Women During Prenatal Diagnosis: Personal Perspectives from the United Kingdom** 83
Helen Statham and Joanie Dimavicius

7   Cooperation Is Rewarding If the Boundary Conditions Fit:
    Interdisciplinary Cooperation in the Context
    of Prenatal Diagnostics 99
    Astrid Riehl-Emde, Anette Bruder, Claudia Pauli-Magnus,
    and Vanessa Sieler

8   Prenatal Genetic Counselling: Reflections on Drawing Policy
    Conclusions from Empirical Findings 109
    Anders Nordgren

9   Taking Risk in Striving for Certainty. Discrepancies in the Moral
    Deliberations of Counsellors and Pregnant Women
    Undergoing PND 121
    László Kovács

10 Ethical Thoughts on Counselling and Accompanying
    Women and Couples Before, During
    and After Prenatal Diagnosis 139
    Dierk Starnitzke

11 Client, Patient, Subject; Whom Should We Treat?
    On the Significance of the Unconscious in Medical Care
    and Counselling 147
    Yair Tzivoni

12 Decision to Know and Decision to Act 155
    Regina Sommer

13 Moral Decision-Making, Narratives
    and Genetic Diagnostics 167
    Göran Collste

14 Prenatal Diagnostics and Ethical Dilemmas in a Mother
    Having a Child with Down Syndrome 177
    Maria Samakouri, Evgenia Tsatalmpasidou, Konstantia Ladopoulou,
    Magdalini Katsikidou, Miltos Livaditis, and Nicolas Tzavaras

15 Is There One Way of Looking at Ethical Dilemmas
    in Different Cultures? 191
    Stephan Hau

Index 205

Contributors

Anette Bruder, social worker, couple and family therapist (SG), member of the scientific staff of the Institute of Psychosomatic Cooperation Research and Family Therapy, University Hospital Heidelberg. Qualitative research and publications within psychosocial counselling in the context of prenatal diagnostics and elderly couples.

Göran Collste, Professor of Applied Ethics, Linköping University, Sweden, whose research and teaching deals with problems in ethics and applied ethics and whose publications include books and articles on the principle of human dignity, work ethics, global justice, and ethical issues related to information and communication technologies (ICT-ethics).

Joanie Dimavicius is an independent health and charitable sector consultant and trainer who worked in the health sector, local government, and charities for over 30 years. She has extensive experience of establishing charities and projects to meet the needs of children, families, and specific communities, was the founder Director of ARC – Antenatal Results and Choices, formerly known as SATFA (Support Around Termination For Abnormalities) and is currently external advisor to the NHS Fetal Anomaly Screening programme (England).

Tamara Fischmann, Dr. rer. med., is a clinical psychologist, psychoanalyst (IPA, DPV), and a member of the scientific and clinical staff at the Sigmund Freud Institute, Frankfurt. She specializes in psychoanalytic basic research, and is methodological and empirical advisor to the institute.

Stephan Hau, Professor for Clinical Psychology, PhD, Psychoanalyst (IPA, SPF, DPV), teaches at the Department of Psychology at Stockholm University. His main research areas are clinical research as well as basic research in the fields of dream and memory research, psychotherapy research, and social psychology (mass group behaviour).

Elisabeth Hildt, PhD, is a researcher in the Department of Philosophy at the University of Mainz and a member of the Chair for Ethics in the Life Sciences, University of Tübingen. The focus of her research is on the theory and ethics in the life sciences, with particular interest in medical genetics, neurophilosophy and neuroethics.

Magdalini Katsikidou is a medical doctor, a trainee in psychiatry and a doctoral candidate at the Democritus University of Thrace, Greece. Her current research work includes the study of victimization of psychiatric patients and/or people with physical disabilities within the context of transcultural psychiatry.

László Kovács, PhD, is a philosopher at the Chair for Ethics in the Life Sciences of the University of Tübingen. His research concerns issues in clinical ethics and in ethics related to medical technologies with a focus on the beginning of life.

Konstantia Ladopoulou is a child and adolescent psychiatrist, Director of the Children's Mental Health Centre of Athens in the Child Psychiatry Hospital of Attika, and a part-time academic staff member of the School of Social Science in the Open University of Greece. She has a PhD in Behavioural Psychotherapies and an MSc in Health Care Management. She has been trained in infancy and early childhood psychiatry at the University of Geneva, Switzerland, and she specializes in brief mother-infant psychodynamic psychotherapies.

Katrin Luise Läzer, PhD, is a psychologist in psychoanalytic training. She works as a scientific assistant at the Sigmund Freud Institute, Frankfurt, and at the University of Kassel, Germany.

Marianne Leuzinger-Bohleber is a training analyst in the German Psychoanalytical Association, Chair of the Research Subcommittee for Conceptual Research, and a member of the Swiss Psychoanalytical Society. Full Professor for Psychoanalytic Psychology at the University of Kassel, and head Director of the Sigmund Freud Institute, Frankfurt. Her main research fields include epistemology and methods of clinical and empirical research in psychoanalysis, interdisciplinary discourse with embodied cognitive science, and modern German literature. She is the coordinator of EDIG.

Miltos Livaditis is a psychiatrist, Professor in Social Psychiatry at the Democritus University of Thrace, and the Director of the University Psychiatric & Child Psychiatric Departments in the University General Hospital of Alexandroupolis, Greece. Apart from his studies in medicine, Miltos Livaditis has got a degree in Law and Political Sciences. He has been actively involved in the movement of the psychiatric reform in Greece and he is particularly interested in studying the philosophy of mind.

Anders Nordgren is Professor of Bioethics and Director of the Centre of Applied Ethics, Linköping University, Sweden. His research concerns ethical issues raised by biomedicine and biotechnology. His main focus is on medical genetics and animal experimentation.

Claudia Pauli-Magnus, Dr. sc. hum., psychologist, specialized in qualitative research. Member of the scientific staff of the Institute of Psychosomatic Cooperation Research and Family Therapy, University Hospital Heidelberg.

Nicole Pfenning-Meerköttter is a psychologist in psychoanalytic training (DPV). She works as a scientific assistant at the Sigmund Freud Institute, Frankfurt and is doing her PhD within the EDIG Project. Her special research interest is knowledge management in complex research projects.

Astrid Riehl-Emde, Professor of Clinical Psychology at the University of Zurich, and assistant director of the Institute of Psychosomatic Cooperation Research and Family Therapy, University Hospital Heidelberg. Head of model projects in the field of cooperation within prenatal diagnostics.

Maria Samakouri is a psychiatrist and Assistant Professor in Psychiatry at the Democritus University of Thrace and the University Psychiatric Department in the University General Hospital of Alexandroupolis, Greece. She is in charge of the rehabilitation programmes in the catchment area of the University General Hospital of Alexandroupolis. Her main clinical and research interests are in social and community psychiatry.

Vanessa Sieler, psychologist and member of the scientific staff of the Institute of Psychosomatic Cooperation Research and Family Therapy, University Hospital Heidelberg. Within the projects in the context of counselling and prenatal diagnostics she is responsible for quantitative methodological and statistical concerns.

Regina Sommer holds a MA in Philosophy and Romance studies from the Eberhard Karls University Tübingen, Germany. She worked within the EDIG study as a scientific assistant at the Department of Ethics in the Life Sciences, University of Tübingen.

Dierk Starnitzke is chairman of the "Diakonische Stiftung Wittekindshof" in Bad Oeynhausen, Germany, a special organisation for integration, where close to 3,000 people with disabilities are supported by more than 2,500 employees. As pastor of the Protestant Church of Westfalen he is also Professor on the Protestant University of Wuppertal/Bethel. He is specialised on New Testament Studies and on Management of diaconic organisations and is therefore teaching mainly at the Institute for Diaconic Science and Diaconic Management in Bethel, Bielefeld.

Helen Statham is a senior research associate in the Centre for Family Research in Cambridge. Since 1990 she has worked on a number of studies of pregnancy and prenatal testing in routine antenatal care. She is a former Chair and current Trustee of the UK voluntary sector group, Antenatal Result and Choices (formerly SATFA, Support Around Termination For Abnormality), and has worked extensively with parents and professionals at all stages of the antenatal screening process in areas of decision-making, good practice, and support.

Evgenia Tsatalmpasidou is a psychologist. Having completed an MSc in Mental Health Studies at the University of London (King's College), UK, her work experience involves counselling services provision for individuals faced with social exclusion. She has been also working with children in a public diagnostic, assessment, and support centre and in the Centre for Child and Family, SOS Children's Village of Thrace, Greece.

Nicolas Tzavaras is Professor of Psychiatry at the Democritus University of Thrace, Greece. He is a training analyst, member of the German Psychoanalytical Association and president of both the Hellenic Psychiatric Association and the Hellenic Psychoanalytical Society. His main interests are the psychodynamic approach of psychosis and transcultural psychiatry.

Yair Tzivoni has an MA in clinical psychology from the Hebrew University Jerusalem and practices Lacanian Psychoanalysis. He also works as a senior psychologist at the Beer Yacov Mental Health Centre.

Chapter 1
Ethical Dilemmas Due to Prenatal and Genetic Diagnostics. An Interdisciplinary, European Study (EDIG, 2005–2008)*

Marianne Leuzinger-Bohleber

Abstract This chapter offers an overview of the EDIG (Ethical Dilemmas due to Prenatal and Genetic Diagnostics) study, a European wide interdisciplinary study investigating the ethical dilemmas with which women and couples are confronted during and after undergoing prenatal and genetic diagnostics (PND). The societal and institutional context, the research questions, the methodological procedures and the combination of empirical, clinical and interdisciplinary approaches are summarized. The focus of the chapter lies on the clinical-psychoanalytical findings. Three extensive single case studies illustrate the complex and individual personal experiences of the different women and couples during and after PND and coping with its ethical dilemmas. The chapter concludes with a discussion of both protective and risk factors. These may help the medical staff to identify couples with problematic psychic constellations who might need support and help particularly during the process of having to decide on the life or death of their unborn child.

Keywords Ethical dilemma • Prenatal and genetic diagnostics • Protective and risk factors digesting an interruption of pregnancy • Psychoanalytic interviews and crisis intervention

1.1 Introduction

As the coordinator of the EDIG study (Ethical Dilemmas due to Prenatal and Genetic Diagnostics) I would firstly like to present an overview of the scientific and societal context of this interdisciplinary study and secondly some of the clinical

---

*Parts of this chapter were originally published in *The Janus Face of Prenatal Diagnostics: A European Study Bridging Ethics, Psychoanalysis, and Medicine*, Marianne Leuzinger-Bohleber, John Tsiantis, Eve-Marie Engels (eds) originally published by Karnac Books in 2008, reprinted with kind permission of Karnac Books.

M. Leuzinger-Bohleber (✉)
Sigmund-Freud-Institut, Myliusstraße 20, Frankfurt 60323, Germany
e-mail: m.leuzinger-bohleber@sigmund-freud-institut.de

psychoanalytical findings from interviews with women and couples during and after prenatal and genetic diagnostics. Because many of the participants of the study are writing their own contributions in this volume, I am not going into the interdisciplinary and empirical findings of EDIG.

Achievements in genetic research produce ethical and moral dilemmas, which need to be the subject of reflection and debate in modern societies. Moral dilemmas are situations in which a person has a strong moral obligation to choose each of two courses of action, but cannot fulfil both. Denial of ambivalences that arouse moral dilemmas constitutes a threat to societies as well as to individual persons. The EU wide study "Ethical Dilemmas Due to Prenatal and Genetic Diagnostics" (016716-EDIG), was undertaken between 2005 and 2008, to investigate these dilemmas in detail in a field which seems particularly challenging: prenatal diagnostics (PND). The existence of PND confronts women and their partners with a variety of moral dilemmas: should they make use of this technique at the risk of hurting the fetus by the technique itself or of being possibly confronted with the decision about termination of pregnancy? Once they have undergone PND, women and their partners may receive information about abnormalities. They must then confront moral dilemmas regarding: the decision about the life or death of the unborn child; the responsibility for the unborn child; responsibility for its well being when born with abnormalities and for its possible suffering. An important aspect is the conflict between individual beliefs and obligations and those of society's specific cultures. These dilemmas have not received full attention in our societies and often remain latent, creating a source of distress for women (and partners) which may be a burden on relationships. Some couples show better coping capabilities, particularly if support from competent professionals is available. However, further research was needed to identify those with vulnerability to psychopathology as a consequence of abortion after PND results or of giving birth to severely handicapped children. Pathology sometimes does not appear until years after the decision. Our study was a contribution to this necessary research.

The study described existing care systems across participating centres in Germany, Greece, Israel, Italy and Sweden. Data was collected in two substudies, with all results integrated into a discourse on ethical dilemmas. Substudy (A) recruited two groups of women with partners where possible (with either positive or negative prenatal diagnosis results. Overall, 1,832 individuals joined the study). Experiences with PND and connected dilemmas were explored using questionnaires and interviews. Results have been discussed in interdisciplinary research groups. Substudy (B) interviewed psychoanalysts and their long-term patients who had shown severe psychopathologies as reactions to the dilemmas mentioned. Results of the study can be used to discuss possible protective and risk factors for women/couples undergoing PND and to give perspectives for training. The EDIG study offered a unique opportunity for multidisciplinary dialogue between ethicists, psychoanalysts, medical doctors, philosophers and cultural anthropologists. Furthermore, the relatively detailed interviews with women/couples after PND as well as the empirical findings from questionnaire data could be used by all the different authors with different disciplinary and cultural perspectives, as was illustrated in our first publication (Leuzinger-Bohleber et al. 2008a, 2008b). This volume will deepen this impression.

1.2 Prenatal and Genetic Diagnostics – New Sources of Human and Ethical Dilemmas

Coping with modern technology in the life sciences and medicine (genetics, biotechnology, atomic physics etc.) has become a major issue for people living in the twenty-first century creating new chances and possibilities but also new dangers and ethical concerns. Probably the best-known example is the discourse on nuclear physics in which the ambivalence of techniques becomes visible. Techniques can be misused in building the atomic bomb or applied for peaceful purposes, producing electricity in nuclear power stations.

In the EU study summarized in this chapter we have investigated ethical dilemmas connected to another of these new technologies, prenatal and genetic diagnostics. This is a relevant discipline in modern life sciences, which seems particularly challenging. In many European countries amniocentesis, for example, is a routine diagnostic tool for women becoming pregnant over the age of 35 years. In the last decades enormous progress has been made in prenatal diagnosis of genetically based diseases and other serious abnormalities. However, several problems have emerged which wait to be answered. This is especially true with respect to ethical dilemmas emerging in this context. We know that a positive prenatal genetic diagnosis creates distress for all women and their partners as they face the difficult decision about the life or death of their unborn child. However, the reactions of women and their partners to this situation vary. Although several individual, religious and social factors and different coping strategies have been described in the literature, further research was needed taking also into account different national and cultural factors within the EU research area. The EDIG study was a first contribution to this discussion.

Identification of groups of women, their partners or specific partnership constellations, who will react with extensive feelings of vulnerability to findings concerning severe abnormalities of the fetus is difficult. According to clinical psychoanalytic experiences and studies, severe depressions or other psychopathological symptoms may be expected as reactions sometimes years after the termination of the pregnancy or its continuation and the birth of a severely handicapped child. However, some women do not suffer long-term adverse effects after termination or this stressful life situation. Support from the partner, the family and professionals seem to be protective factors as may be more open discourses on ethical dilemmas in society. This study aimed to investigate and describe both protective factors and risk factors. Individuals or couples who are particularly sensitive should get special support during the crisis due to positive prenatal findings.

Nevertheless, how women/couples react to this situation may depend on how this ethical dilemma is addressed by professionals working in the healthcare system. This important issue was another focus of the two substudies of the project A and B (see below).

The objectives of the project addressed the problems and dilemmas due to technological progress in science and technology from the individual perspective of the women undergoing the process of prenatal diagnostics, their partners and the health professionals working with them. These objectives are highly relevant for bridging the gap between ethics and developments in the life sciences and their technologies. It focused on promoting the interdisciplinary exchange with ethicists and clinicians who are experts in uncovering possible consequences for psychic health of negotiating ethical conflicts. This primary objective was achieved by creating an exchange between the experts of various disciplines relevant for these issues, including ethicists, philosophers, clinical psychologists, psychoanalysts and medical doctors working in this field. Whereas the cooperation between ethicists and medical doctors is already well established in the field of medical ethics, the intensive exchange between ethicists and clinical psychologists as well as psychoanalysts, as took place within the EDIG study is rather innovative. Another innovation was that highly theoretical discourses in ethics and philosophy have been connected with both detailed clinical observations from interviews with couples during and after decision making around PND and empirical data from questionnaires. This also allowed discussion of one of the most important questions of modern ethics, namely the relationship between empirical findings and ethical principles and standards. Large differences in the perception, assessment and ways of dealing with these ethical conflicts exist in different European countries (see e.g. Nationaler Ethikrat 2003). For this reason countries with different cultures and traditions were part of the study, including: Sweden with a long liberal tradition about abortion; Greece, a country where some parts have a more profound religious tradition (both majority: Christians and 1% Muslims); Israel and Germany, because the number of terminations after problematic prenatal and genetic findings is extremely different between the two countries (80% in Israel and 23% in Germany) (cf. Raz 2004); and Italy, a country with a catholic background in a changing societal situation, for example the dramatic decrease in the number of children per couple). Thus, the project addressed culture specific factors and generated original findings on these relevant issues. In addition, our study contributed to deepening the understanding for the cultural differences within the EU as well as to be the first to establish shared quality standards in this field. This is an important step in the process of the culturally sensitive integration of European differences.

The most important aim of this study was to address major ethical dilemmas in the context of genetic and prenatal diagnosis, and through that help to create greater awareness of, and sensitivity to these dilemmas among various segments of the societies. Another important contribution was to develop strategies and therapeutic tools for women and their partners to help them deal with the problematic consequences of their situation. A multidisciplinary approach helped to involve various healthcare providers, to improve knowledge transfer and to support dissemination. The results also are relevant for political decision makers, in the different European countries (as members of the national ethic committees), which have been or still will be included and addressed in the dissemination of the findings of our study (see www.sigmund-freud-institut.de).

1.3 State of Art of Knowledge in the Area of Research as Starting Points of EDIG[1]

1.3.1 Knowledge Based on the Empirical Studies on Decision Making After PND

Our study was a continuation and in parts a transcultural replication study of the investigations of a research team in Cambridge (Statham et al. 2001). Statham (2002) summarized the literature on women's experiences of PND as follows:

> "In the absence of prenatal screening, some 2% of babies will be born with a structural anomaly, in 7 out of 800 with Down syndrome, and similar numbers with chromosomal or serious genetic disorders. Most of these occur in healthy women who enter pregnancy with little thought to being 'at risk'. In most countries, prenatal screening for such fetal abnormalities will be offered as a routine component of the antenatal care for women. If a diagnosis is made, decisions have to be made about management of the pregnancy. The nature of the abnormality, the gestation at diagnosis and the prevailing legal framework for abortion will determine the options available to parents.
>
> Research that investigates decision-making after prenatal diagnosis and parents' psychological response after termination or continuing the pregnancy have been recently reviewed. The principle findings are:
>
> - When given a diagnosis parents are shocked and write of 'reeling from shock, numbed by a sudden catastrophe' (Brown 1989) or 'the image of myself, alone, screaming into a white plastic telephone' (Rapp 1984).
> - It is widely accepted that decisions should be 'informed', an informed decision being one in which values, knowledge and attitudes, and behaviour of the decision-maker are congruent (Marteau et al. 2001). However there is little research evidence about how women make decisions following the diagnosis of a fetal abnormality.
> - Whatever the abnormality and whatever the decision they subsequently make, parents will lose what they had believed previously was a normal pregnancy (Green et al. 1992; Marteau and Mansfield 1998; Statham 2003). Experiencing symptoms of acute grief including anger, despair, guilt, inadequacy, sleeping and eating difficulties (Hunfeld et al. 1993) they have no choice but to make difficult decisions about the management and outcome of the pregnancy.
> - After a termination virtually all women show an acute grief response and psychological distress is high in the short-term (Hunfeld 1995; Hunfeld et al. 1993; Elder and Laurence 1991; Black 1989; Zeanah et al. 1993; Iles and Gath 1993; Salvesen et al. 1997). For example, in the study by Iles (Iles and Gath 1993) 41% of study participants showed symptoms of psychiatric morbidity when assessed using the Present State Examination four to six weeks post-termination. Psychological distress declines over time for most women (maximum follow-up time of four years) (Hunfeld 1995; Hunfeld et al. 1993; White-van Mourik et al. 1992; Black 1989; Iles and Gath 1993; Salvesen et al. 1997) although some remain distressed up to 2 years after the termination.
> - A number of factors have been investigated as potentially predictive of lasting psychological distress. They can be classified into three main categories: the obstetric

---

[1] State of art concerning the discussions in modern bioethics: see Engels and Hildt in Leuzinger-Bohleber et al. 2008b.

characteristics of the termination, socio-demographic characteristics of the woman and her family, and characteristics of the care given to the mother. Findings across studies are variable and as with other perinatal losses, little can be said with certainty beyond the importance of social support (Zeanah 1989; Gregg 1988; Hunfeld 1995; Chambers and Chan 2000; and in Toedter et al. 2002; Lasker and Toedter 2000).
- Little attention has been given until recently about parental needs when continuing a pregnancy after a prenatal diagnosis when care focuses on the baby.

It has been suggested that these inconsistencies have arisen because methodologies are poor. An alternative to the view that existing studies are not good enough is that perinatal grief is too complex an emotion to seek to quantify. Recent research on grief and bereavement is challenging earlier models, in particular those that assume that the absence of a complicated process of grieving is pathological (Stroebe 1992). While earlier models emphasised the 'sameness' in individuals' responses to bereavement, the individuality of the response to grief is increasingly being identified, challenging the assumption that there are universal stages of emotional adjustment and universal tasks of grieving that have to be completed (Neimeyer 1999). Rather than measuring features of grief such as 'core bereavement' (Burnett et al. 1997) or 'maladaptive symptoms' (Prigerson et al. 1995) it is possible to use different strategies such as understanding individuals' narratives to explore the personal impact of the pregnancy loss (Neimeyer 1999)". (Statham 2002, p. 236).

Thus taking into account the limited empirical research on decision making after PND, the EDIG study has been one of the largest empirical studies to have been performed. Specific features of the study are that it is EU wide, is prospective and combines quantitative and qualitative approaches, a multidisciplinary theoretical perspective and a practical orientation.

1.3.2 *Psychoanalytical and Clinical Knowledge*

Understanding individual narratives and their idiosyncratic conscious and unconscious experiences belong to the aims of a psychoanalytic treatment. Thus, we were convinced that the specific psychoanalytic knowledge based on the intensive therapeutic experiences with pregnant women as well as with women or men who had gone through dilemmas due to PND could add an important and unique dimension to the multidisciplinary dialogue on these issues. The data gathered in long-term therapies with individuals cannot be collected by any other research method. Therefore, we wanted to systematically investigate the clinical observations of psychoanalysts and former patients concerning different coping possibilities with dilemmas due to PND.[2]

Many clinical psychoanalytic studies have shown that pregnancy is a central event in women's life cycle, confronting them with the value and potentials as well as with the anxieties of women's gender identity (see e.g. Giovacchini 1979; Pines 1993; Stern 1995; Bassin et al. 1996; Leuzinger-Bohleber 2001).

---

[2]The resources of EDIG only allowed us to realize this part of the study, Substudy B, in Germany, Greece and Sweden.

For example, pregnancy enables a process of integrating infantile fantasies on ones female body and identity into a self-concept in which creativity as a female individual may be combined with motherhood. On the one hand this process means a new opportunity in the life of women (and their partners). On the other hand, this may also be a source of danger for one's own development. Pregnancy may bring about unconscious fear of retaliatory attacks on one's inner space and one's fertility or confront women with their unconscious fantasies and impulses which often have a quality which psychoanalysts characterize as "archaic". According to psychoanalytic theories the origin of these archaic fantasies can be found in real experiences in early childhood and childhood fantasizing, psychic processes in periods of the human development, which have been dominated by relatively primitive ways of thinking and feeling. According to a "simple" attempt for orientation this state of mind is "pre-ambivalent": The psychic realities are rigidly split into absolute categories (differentiating between "good" and "bad", "right" and "wrong" etc. see below). But because the human psyche does not seem to forget anything, these archaic modalities of psychic functioning have been perceived in the unconscious. In everyday lives of "normal adults" these archaic states of the mind are not mainly influencing current emotional and cognitive processes because they are well integrated in more mature modalities of psychic functioning and kept in the unconscious by functional defence processes. However: in extreme (traumatic) situations, with some structural analogies to the original infantile experiences, these unconscious fantasies can be reactivated and then determine the current psychic realities. We think that the traumatic situation after a positive finding in PND is structurally predetermined to reactivate early experiences on life and death, which have been kept in an archaic dimension of one's unconscious.[3]

To mention just one example a little bit more in detail: Leuzinger-Bohleber (2002, 2006) discussed her clinical observations in long-term psychoanalysis with women who had decided to interrupt their pregnancy. Some of these women suffered from severe unconscious guilt feelings even years after the interruption. These guilt feelings were connected to the just mentioned archaic level of psychic functioning, an unconscious "truth" to be a "totally evil, condemnable child murderer" who never deserves to be happy again and has to fear revenge. These fantasies had been one of the unconscious sources of severe chronic depressions as well as of sleep disturbances, anxieties and psychosomatic symptoms. They were connected to an ubiquitous female fantasy system, which Leuzinger-Bohleber (2001) described as "Medea fantasy".[4] The "Medea fantasy" may, as Freud (1908) presumed, have been an early-infantile day-dream fantasy of the

---

[3] By the way: this psychoanalytic thesis gets new support by contemporary neurosciences, which also postulates that "the brain does not forget anything…" (see Leuzinger-Bohleber et al. 2008a, 2008b).

[4] The unconscious fantasy is probably still the most central psychoanalytic concept. Although other therapeutic disciplines, academic psychologists, cognitive and neuro-scientists talk about 'unconscious information processing', too, they usually mean a different, a *descriptive* unconscious. Above all the *concept of the dynamic unconscious* which assumes that human behaviour is unconsciously determined by repetition compulsion still serves to distinguish psychoanalysis from the outside and to assure common ground from the inside.

women, in which earliest bodily experiences and primal fantasies (as e.g. on the primal scene, birth and death etc.) had been included and, as Sandler and Sandler (1983) postulated, probably had been banned into the unconscious in the 4th or 5th year of life, becoming part of the dynamic unconscious. The content of the fantasy may have been rewritten 'nachträglich' again and again, e.g. by gender fantasies in adolescence as well as fantasies on motherhood and femininity in late adolescence. In the centre of this unconscious fantasy was the conviction that female sexuality is connected with the experience of existential dependency on the love partner and the danger of being left and narcissistically hurt by him. The women unconsciously fear their own sexual passion might revive uncontrollable destructive impulses in a close, intimate relationship, which are directed against the autonomous self, the love partner and above all against the offspring of the relationship with him, against their own children. Thus they are unconsciously convinced to be capable to kill their own children in a situation of extreme dependency, devaluation and narcissistic injury (see myths of Medea by Euripides).

One of our theses in EDIG was, that this "archaic" unconscious fantasy system and similar ones can be reactivated in a situation like PND where a woman has to decide on life and death of her unborn child, psychoanalytic hypotheses which we tried to clinically evaluate further in EDIG (see Leuzinger-Bohleber et al. 2008b, p. 151 ff.).

In this context it also seems important to consider the fact that PND takes place during a period of pregnancy in which many women develop an intensive attachment to their unborn child which can be observed e.g. in specific fantasies, dreams or daydreams of these women. Therefore having to interrupt pregnancy and thus losing the unborn child might not only arise intensive feelings of loss and mourning but also severe guilt feelings. Seen from a psychoanalytic perspective the situation, of having to "kill" one's fetus, thus easily can reactivate the unconscious "Medea fantasy" or other primitive unconscious fantasies. If these reactivations are not recognized and "worked through" the woman will be determined by the reactivated primitive ("archaic") level of psychic functioning which might become one of the possible sources for not being able to overcome the psychic crisis after the interruption within some months but developing a yearlong psychosomatic or psychic illness. From a psychoanalytic perspective we suppose that such psychodynamic factors could be responsible for the minority of women, mentioned above who seem to suffer from long-lasting psychosomatic and psychic symptoms after an interruption of their pregnancy.

To mention just another example: For some women PND seems unconsciously experienced as a possibility to "prove" that the baby was not harmed by one's own destructiveness. According to Langer and Reinhold (1993) for many couples negative PND findings are indeed a relief and thus a chance for a less burdened pregnancy. In contradiction to this, Hohenstein (1998) has interviewed 14 pregnant women with similar motivation for PND. Hohenstein found that their anxieties did not decrease when they received negative findings but rather increased. Nevertheless, all these women would undergo PND again but pled for a supporting professional attitude of the medical staff including the ethical dimensions in the context of decisions after PND. The results yield in the necessary condition of offering counselling possibilities for couples after PND (cf. Braehler and Meyer 1991). Ringler and Langer (1991) developed an interdisciplinary counselling concept for couples with the diagnoses of severe "fetal abnormalities".

Many psychoanalysts have reported on their clinical experiences that some women suffered from severe depressions often after an abortion or after the birth of a severely handicapped child, despite of positive prenatal findings. As just mentioned, depressions were seen to be due to unbearable unconscious guilt feelings (Leuzinger-Bohleber 2000, 2002, 2006). Girardon-Petitcolin (2001) describes how the painful choice of the impossibility and traumatic separation from a malformed fetus seems to summon up the parts that are not represented in oneself. This process can be seen as condensed in the figure of the "dead child", summarizing the failures in the subject's history of symbolization. It imposes the fall of the wonderful child of primary narcissism and creates a void that attracts "melancholiform" reactions in which the subject may be afraid of losing herself. In contrast, the opportunity for elaboration within the clinical setting offered through these situations allows – in the best of all cases – a chance for what has not been symbolized to be taken up once more and open the possibility for exploration of other subjective potentials.

It is interesting that not all women or men in therapy who had undergone a termination of pregnancy reacted with severe psychopathological symptoms. Personal, idiosyncratic, resilient individuals, familiar, and social supports seem to be protective factors as well as a public atmosphere that allows open discussion of the ethical dilemmas. Such open discourse seems to help those couples who are in the situation of having to decide over life or death of an abnormal fetus (and other related ethical issues) to experience grief in a direct way. Thus, they might be able to talk openly about the difficult emotions and fantasies involved instead of having to taboo them. Seen from a psychoanalytic perspective, the latter case increases the risk to ban emotions, conflicts and fantasies into the unconscious and consequently to an increased risk for developing psychopathological symptoms at some point later in life. It also increases the probability for a regression onto the archaic level of psychic functioning mentioned above.

In EDIG we further studied these clinical hypotheses. The multidisciplinary framework of the psychoanalytical interviews offered a unique possibility to compare psychoanalytic and non-psychoanalytic understandings of the clinical findings as will be illustrated in the different chapters written by the experts in ethics in this volume.

1.4 Participants in EDIG

Table 1.1 gives an overview of the participating scientists in EDIG, state 2008 (remark: some of the dates have changed in the meantime).

Table 1.1 Overview of the participating scientists in EDIG

Participant name	Country	Participating institution
Marianne Leuzinger-Bohleber (Prof. Dr. phil., clinical psychologist, psychoanalyst)	Germany	University of Kassel/ Sigmund-Freud-Institut
Manfred E. Beutel (Prof. Dr. med., psychiatrist, psychologist, psychoanalyst)	Germany	University of Mainz/SFI
Antje Haselbacher, (Dr. rer. medic, psychologist)	Germany	University of Mainz
Tamara Fischmann (Dr. rer. med., Dipl. Psych., clinical psychologist, psychoanalyst, methodological expert for the project)	Germany	Sigmund-Freud-Institut
Nicole Pfenning (psychologist)	Germany	Sigmund-Freud-Institut
Katrin Luise Läzer (Dr. phil.)	Germany	Sigmund-Freud-Institut
Samuel Fischmann (Dr. med., Gynaecologist)	Germany	Sigmund-Freud-Institut
Ulrike Beudt (Dr.med., genetic scientist)	Germany	Sigmund-Freud-Institut
Bernhard Rüger (Prof. Dr. rer. nat., statistician)	Germany	University of Munich/SFI
Dagmar von Hoff (Prof. Dr., expert in cultural, gender and literary studies)	Germany	University of Mainz/SFI
Axel Scharfenberg (Oeconomist)	Germany	Sigmund-Freud-Institute
Eve-Marie Engels (Prof. Dr. phil., Chair for Ethics in the Life Sciences, Faculty of Biology, and Interdepartmental Centre for Ethics in the Sciences and Humanities, IZEW, Spokeswoman)	Germany	Eberhard Karls University of Tübingen
Elisabeth Hildt (PD Dr. rer. nat., Chair for Ethics in the Life Sciences, Faculty of Biology, and Interdepartmental Centre for Ethics in the Sciences and Humanities, IZEW)	Germany	Eberhard Karls University of Tübingen
László Kovács (Dr. phil., Chair for Ethics in the Life Sciences, Faculty of Biology)	Germany	Eberhard Karls University of Tübingen
Regina Sommer (M.A., Chair for Ethics in the Life Sciences, Faculty of Biology)	Germany	Eberhard Karls University of Tübingen
Rachel Blass (Prof. Dr. phil., clinical psychologist)	Israel	Hebrew University of Jerusalem
Vardiella Meiner (M.D., geneticist)	Israel	Hadassah medical organization, Hebrew University hospital
Nili Yanai (M.D., gynaecologist)	Israel	Hadassah medical organization, Hebrew University hospital
Yotam Benziman (PhD, philosophy, ethicist)	Israel	Hebrew University of Jerusalem
Gideon Bach (Prof. Dr. med., Head of Human Genetics)	Israel	Hebrew University of Jerusalem
Beatriz Priel (Prof. Dr. phil., Head of Clinical Psychology, Track: Behavioral Sciences)	Israel	Ben Gurion University of the Negev
Rivka Carmi (Prof. Dr. med., Head of the Genetics Institute)	Israel	Ben Gurion University of the Negev
Nicolas Tzavaras, MD Professor of Psychiatry – psychoanalyst	Greece	Democritus University of Thrace

(continued)

1   Ethical Dilemmas Due to Prenatal and Genetic Diagnostics

Table 1.1 (continued)

Participant name	Country	Participating institution
Maria Samakouri, MD Assistant Professor of Psychiatry	Greece	Democritus University of Thrace
Konstantia Ladopoulou, MD Lecturer in Child Psychiatry	Greece	Democritus University of Thrace
Evgenia Tsatalmpasidou, MSc, psychologist	Greece	Democritus University of Thrace
Katsikidou Magda, MD	Greece	Democritus University of Thrace
Miltiadis Livaditis, MD Associate Professor of Social Psychiatry	Greece	Democritus University of Thrace
John Tsiantis (Prof. Dr.med., psychiatrist)	Greece	Athens University
Stephan Hau (PD.Dr.phil., clinical psychologist, psychoanalyst)	Sweden/ Germany	Linköping University/SFI
Göran Collste (Prof. Dr. theol., Center for Applied Ethics)	Sweden	Linköping University
Anders Nordgren (Prof. Dr.phil., Center for Applied Ethics)	Sweden	Linköping University
Helen Statham (psychologist, Senior Research Associate, Centre for Family Research)	England	University of Cambridge
Joanie Dimavicius (Consultant in Health and Charitable Sector)	England	University of Cambridge
Gherardo Amadei (Prof. Dr. med., psychiatrist, psychoanalyst)	Italy	Università Cattolica del Sacro Cuore di Milano
Ilaria Bianchi (PhD, psychologist)	Italy	Università Cattolica del Sacro Cuore di Milano
Diletta Fiandaca (psychologist, psychotherapist)	Italy	Università Cattolica del Sacro Cuore di Milano
Edgardo Caverzasi (Prof. Dr.med., psychiatrist, psychoanalyst)	Italy	Università di Pavia DSSAP CIRDIP, IRCCS S. Matteo Pavia
Filippo Sarchi (Dr.med, psychiatrist)	Italy	Università di Pavia DSSAP CIRDIP
Irene Cirillo (Dr.med, psychiatrist)	Italy	Università di Pavia DSSAP CIRDIP
Prof. Eugenijus Gefenas	Lithuania	University of Lithuania

1.5   Summary of Some Clinical Findings of EDIG and Selected Interviews

In the above-mentioned publication (Leuzinger-Bohleber et al. 2008b) we summarized seven selected interviews with women/couples after prenatal and genetic diagnostics as well as the procedure for performing and analyzing the interviews. All the interviews have been evaluated by a group of experienced psychoanalysts ("expert validation") in order to guarantee that the interviews were summarized and condensed in a professional way. The group also helped to select those seven interviews, which seemed best to illustrate the broad range of possible coping strategies with ethical dilemmas by different women and their partners. The narratively summarized interviews have been a central supplement to the findings of the quantitative parts of EDIG (see Fischmann in this volume) because they offer a detailed impression of the individual experiences that each of these women and their partners have gone through, what PND meant to them, particularly if they were unexpectedly confronted with positive findings, how they tried to decide in this situation, what ethical dilemma they mentioned and if and how they felt supported or not in their decision and experiences etc. A unique feature of the EDIG study was that these interviews could be used as joint information for the multidisciplinary exchange. All the authors of our first publication referred to the same interviews (see Leuzinger-Bohleber et al. 2008b).

The summaries of the interviews could not be shortened too much because they need to illustrate the complexity as well as the individuality of the experiences that the specific woman and her partner are going through after PND. In our earlier publication the seven interviews represented a broad spectrum of different situations: a decision against PND (Interview 1, interview 7), situations after having interrupted pregnancy due to positive findings (interview 2, 3, 4, 6) and the decision to continue pregnancy in spite of a trisomy 21 (interview 5). In the limited space of this chapter we can just reprint two of these interviews and one of substudy B.

Interview 3: "The child without a face...."

This interview from substudy A was chosen in order to illustrate how a couple may face a situation of deciding on pregnancy interruption even if they did not choose to use PND.

Mrs. C. immediately called back after her therapist told her about our research. She and her husband are willing to come to a hotel for the interview. In my eyes they are an attractive couple, both tall persons and slim. Both are academics working in different areas. While Mrs. C. is in the bathroom, casual small talk arises between her husband and myself. He talks about his work and the fact that he spends much time travelling. After a short pause, I once more explain the target of our survey. Both listen attentively. A long and coherent report by Mrs. C. follows, calm but still emotionally involved. Her husband supports this report attentively, in a discreet and sensitive manner.

Mrs. C interrupted her pregnancy 5 weeks previously. Although she is nearly 40 years old and her husband even a little older, they decided against amniocentesis, as they were very aware about the dilemma they would become involved in if receiving positive diagnostic findings. Additionally they did not want to take the risk of a spontaneous abort, which happens in one in 200 cases. Both report that – in contrast to many others – they know what a probability of 0.5% means (Mrs. C. too, is a natural scientist). They were concerned about the fact that, at their age, the probability of giving birth to a disabled child is increased. Nevertheless they assumed that "everything is okay", because Mrs. C. felt very well during the first weeks of her pregnancy. But soon they were confronted with a shock: During a routine ultrasonic testing the doctor told her that something was wrong: The fetus, so the doctor, had no face. They were referred to a further special ultrasonic testing: the diagnosis was confirmed. It was horrifying for the couple. In particular both found it dreadful that the child was shown to them on a giant screen by the doctor – without a face. "Here you can see its ears; where the face should be, there is nothing..." Due to the enlargement it looked awful. This happened on a Thursday – on Wednesday she had been given information about the complications for the very first time. On Saturday she had the appointment for the interruption of her pregnancy. Therefore Mrs. C. was asked to start intake of the tablets at once: there was hardly any time for reflection or for valediction. On Thursday she read a book written by a fellow sufferer "Expecting a baby, sudden end", who had been in the same situation – which was very helpful for her. Nevertheless the whole experience was a "traumatic situation" (Mrs. C.): both of them were so pleased about the pregnancy. Both often state that they would have felt happier if the doctor had not noticed the abnormality on her simpler apparatus, and the consequence had been a spontaneous abort. "Admittedly, the child might then have been bigger, and the risk for mother and child might then have been higher, etc....." "In this respect, one has to be thankful for the technical possibilities…" Mr. C. adds. Yet, in terms of the childbirth, Mrs. C. was lucky, it took only 5 h, and due to the painkillers was endurable. They told her that childbirth could take up to 36 h. This was a horrid thought for her. Presumably it was positive that the child was not so big at this time. She herself did not want to see the child. Mr. C took a look. He showed me the extent of his distress when telling me that he did not feel sure that looking at the child had been good for him. Now he has got a picture in his mind, which does not disappear.[5] We discuss, that the confrontation may prove helpful for some people, because reality is often less awful than fantasy. Mr. C seems to doubt this. But still he could – should Mrs. C. ever want it – tell her about his perception. Additionally, a photo still exists, although only a Polaroid photo. They have not yet taken a look at it. It is still enclosed in a sealed envelope. This, by the way, is a proposal: there should be better cameras at the clinic's disposal. Further the room was unfriendly and dowdy. Mrs. C. had wished the room to be painted in friendly colours, with a bunch of flowers inside etc. In addition it felt painful to hear the noises of childbirth and the crying of the newborns inside the maternity ward, and her child was dead. There is definite scope for changes: institution, staff and premises should be committed to the special situation of the couples in a more appropriate way.

For Mrs. C. realising what had happened was worst when she got back home. Mr. C. invited her to join him on the ship to participate in an expedition to distract her, a strategy which subsequently turned out to be totally illusory. Physically, as well as emotionally she was not fit for it: the couple indicates that it came to a severe crisis, because he started working again, and only gave up his travelling plans, when he saw what a bad state his wife was in. He showed empathy in emphasising that the partner's release from work for 1 week should become institutionalised in such cases – this would document that it is "not merely a stronger menstrual period, but the loss of a child…" Both had not paid enough attention to this fact. "And I was in an awfully bad state for a whole week…" Mrs. C. reports.

It proves of great help for Mrs. C., who had lost her brother from cancer a few months before, that she has therapeutic conversations with me so that the two experiences of loss do not mingle too much. It is perceptible that she latently accuses her parents for being so engaged in mourning over their son, that there is no room for mourning over their dead granddaughter and that there is hardly any empathy for their daughter's traumatic experience.

Many times, Mr. C. tries to comfort his wife during the interview, telling her that she could get pregnant again – and then hopefully everything would be alright. Both seem to fear a second disability. They await the results of the pathological examination. They have no genetic dispositions in their families, she neither smokes nor drinks – but nevertheless feelings of guilt arise in such a situation, and they ask themselves what they had done wrong. However both think that the deformity could have hearkened back to a spontaneous mutation.

Mrs. C. would – now that she finally is able to "work again a little", would like to write a report in a journal referring to this topic – a quasi way of coming to terms with her loss.

---

[5]There is a big discussion in the UK currently about seeing babies – new NICE guidelines discourage professionals from encouraging it, though most bereavement support organisations are not in support (Commentary: Helen Statham).

I express my wish that she will get the chance to deal with less distressing topics during a normal pregnancy. "I am a person that can abide the good and the bad beside each other..."

She would like to be in touch with our survey and possibly use this as an opportunity to disseminate her report during the course of our congress. With the help of her report, she wants to help couples that are in similar situation – and this way they hope to contribute to an improvement of the professional handling of abortions connected to prenatal diagnostics. For example, dealing with the dead child was difficult: after several phone calls, they were informed that the child had still not been transferred to the pathology department, but still "was lying about somewhere..., we passed the hospital by day and by night without knowing if our dead daughter was there..." They did not want an individual funeral for their daughter, but are ambivalent about a collective funeral, which will take place months later. Both request more frequent dates for those funerals. To me it seems as if both fear the situation of participating in this ritual together with other couples who they do not know.

At the end of the interview they ask whether, in case of a further pregnancy, it is advisable to consult the same hospital. On the one hand the premises will remind her of the traumatic experience, on the other hand she already knows the staff and they know her and her history. I advise her to discuss this with Mrs. M., her psychotherapist.

Summary of the experts' validation:

Within the group initial associations revolve around thoughts like "technical development makes 'murderers' out of parents"; "how can one endure being pregnant with the certainty that the child has no face". From that moment when technology had been implemented, the further course of the pregnancy seems to have been predetermined.

The question arises why there was no free rein given to the situation; why they did not allow natural childbirth with subsequent death, in order to be saved from acting as 'murderer'. The "haste" is rooted, presumably, in the fact that the idea of carrying a faceless child is not endurable. Had there been no intervention, as Mr. C. points out, his wife's life might have been in danger. Was this medical fact, or his own terrifying fantasy? The hasty interruption could then be seen in this context. Massive feelings of guilt seem to play a role as well in the interview, so it must be assumed: "Are we guilty of the deformation of our child, because we have waited for such a long time and are now too old?"

Two foci emerged in the expert group's discussion:

1. The experience of becoming pregnant intentionally and the pleasure, joy and pride of expecting a baby is abruptly disrupted by the doctor's words: "The fetus has no face". For the couple an abortion seems horrifying because it means destroying the physical and psychological bond between parents and child. The topic of killing a handicapped baby is associated with this situation.
2. The individual dynamics between the couple could have determined the way the situation is handled: the terrible hurry, the husband's fantasy that everyday could continue since the interruption would just be something "like a strong menstruation". Could Mrs. C.'s breakdown have been avoided if someone had taken over the holding function during the acute crisis of decision and interruption? (Even reading a book seemed to help her a bit at first).

The group discusses possible fantasies about "the break down": The abortion is probably experienced as a place where something 'terrifying'/'dirty' is ejected (the child 'without a face' which had to be removed because it might endanger the mother's body may have evoked anal fantasies). Afterwards the couple seem to be scared about this thought. Does Mrs. C. try to express these experiences when she accuses the staff: "It could have been handled differently, the room could have been more colourful, flowers in the room, etc."

At the end of the discussion the topic of the individual funeral arises. Against the background of the couple's history, it is pointed out again that "children who could not be brought into the world are treated in a dishonourable way". Due to the collective funeral a massive process of dispossession, as well as dehumanisation takes place. Funerals should be held contemporary to the death, in terms of funerals of our dead ones, as befits the first cultural achievement of mankind.

One "motto" of this interview was described: Due to modern prenatal diagnostics everybody can be confronted with an unexpected 'discovery' of a severe abnormality of the fetus which entails a situation in which one has to decide on the life and death of an unborn baby. The sudden loss of joy, pleasure, bonding to the growing baby may have a traumatic quality. A late pregnancy interruption is not merely "like a strong menstruation." For couples in this situation a professional 'holding' by staff, friends, the family, psychotherapists etc. might prove very helpful for coping with these burdening experiences.

Interview 5: "Now we have our little toad at home. No, no, seriously we tenderly call her our 'little beetle' because of the way she keeps her legs bent…"

In contrast to the interviews with the couple just mentioned the following couple from substudy A decided not to interrupt pregnancy despite the PND result indicating that their daughter would have 'trisomy 21'. In order to give an impression of their experiences during the phase of decision-making and after the birth of their daughter we summarize here the two follow-up interviews with the mother and the father of the now 6-months-old trisomic child.

Interview with Mrs. E

Mrs. E. who was eager to be interviewed, talked a lot and vividly during the entire interview – as she already had in her first interview with me when she was 5 months pregnant.

She immediately begins to explain, that her daughter was born 6 months ago by Caesarean section, because "she decided to dance around on my sciatic nerve" during the last few weeks of pregnancy. "We are so happy", she continues, "because B. has only a light version of trisomy 21 and practically no comorbidity. The only attendant symptom apparent is a 'ventricle-septum-defect' with three tiny little holes in the cardiac septum. One of them has closed up already on its own and the other two will follow soon", Mrs. E. says reassuringly. "B. was relatively small when she was born, but she is now developing and growing well. She has just started to crawl and babble very sweetly". Mrs E. imagines that B. will develop just like any other baby in her first year and subsequently be outstripped by the others. Mrs. E. is full of joy about her little daughter, something she never imagined she could feel: "It is so wonderful with her, we are so happy and it is such a joy coming home every day. She loves her father, but when I come home from work at night, it is me who is the most important one. Then she is practically living all over me… Papa is a little bit jealous… I am working fulltime again, as we had planned all along, in fact I was called back to work when still on maternity leave. This has been very hard. And I still do have days when I ask myself what I am doing with these silly people at work – why don't I stay home with my little daughter instead… ? We always had agreed that my husband will stay at home for the first 2 years and that I go back to work again (she earns more money than he and he had to change jobs anyway on account of health problems). We have also decided to move into an apartment in order to save money. We already rented a little garden next to our new apartment where our little daughter will be able to play". They are well known in their village, last but not least due to their very colourful baby carriage. "Our daughter has been fully accepted by everyone already...." "Had you been concerned about this?" I ask her. "Yes, particularly regarding elderly people, e.g. our neighbour…but now she is like a grandmother to our daughter and just loves her.... But even if people had not accepted our decision: that would be their problem and not ours! We took a very conscious decision to have our little girl!" I ask if their decision has been questioned. Mrs. E. denies this: she very actively offers information and her viewpoint even where not asked. In contrast to her husband who – after joining a baby crèche where parents meet weekly with their babies aging from newborn up to 3-year-olds – has not offered any information on the trisomy of his daughter to the others, but has decided to wait until the others ask. She is now 6 months and no questions asked so far; he is considering giving some information. "We both would like to prepare our daughter in case she should ever be asked why she is different from other children. She should then say: I just have one thing more than you; namely a chromosome! We hope that she will not make any negative experiences e.g. with other children who sometimes ask awkward questions. Parents of trisomy 21 children have told us that their children had been very disappointed because they were rejected for hugging other children spontaneously...". I point out to Mrs. E. that she wants to make sure that B. will know how to meet any obstacle, e.g. being named 'mongoloid'. She reacts indignantly over that term. In her job she sometimes meets with people from Mongolia and "Trisomy 21 has nothing to do with Mongolia, using this term is purely discriminating!" She wants her daughter to identify with a motto known from an organization for the handicapped: 'It is normal to be different'.

Mrs. E. then continues to elaborate on her experiences in the clinic. She made sure to speak to doctors before being hospitalized to inform everybody there that she will give birth to a trisomy 21 child. Thus everybody was informed and empathized with her situation. They even admired her for her decision. Her daughter had to be checked in the children's clinic which was located in a different building from where she was hospitalized. "The second day after my c-section I made a sitting strike in my wheel-chair at the children's clinic stating that I wanted to have my baby... they finally handed her out to me and allowed me to take her with me. After his initial indignation the doctor eventually even expressed admiration – I still have very good contact with him..." In the last part of the interview Mrs. E. talks again about her painful experiences with the loss of her first two 'babies'. She had two stillbirths within 1 year. B. was her third pregnancy during that 1 year, despite the fact that she is already 40-years-old. "We are so happy to have her ..." On the 'day of remembrance for stillborn children' she went with B. to the grave of her grandmother (who took care of her during her whole childhood). "I stood there remembering my two dead babies and thought to myself 'I cannot even really tell B. that I am here because of her siblings, since if I had had one of them she would not be here'... ". At the end of the interview Mrs. E. wants to know how many trisomy 21 children we have in our study. She is shocked to hear that B. is the only one!

Interview with Mr. E

Mr. E. comes to the interview with 6-months-old B. She sits on his lap during the entire hour of the interview, sucking his thumb or her comforter. He obviously has good contact with her. For me it is obvious that B. is a trisomy 21 child (eyes, tongue, she has difficulty holding her head by herself etc.). I realize that I am irritated that Mr. E. proclaims several times during the interview that other people do not realize that his daughter suffers from trisomy. "We are very, very happy with her – everything is fine..." Mr. E. says at the beginning of the interview. He summarizes their experiences during the 9 months since our last interview. "It was the right decision. I enjoy taking care of her. For my wife it is hard to go to work every day – but she has to cope with this – we always planned it this way. My wife is often jealous that I am the most important person for B. Even if I leave her alone with my wife for a little while she misses me and wants me to come and see her.... B. determines my whole day – she has to be taken care of from 8 a.m. on all day long: I can never sleep longer and sometimes I don't even find time to cook a meal for myself." He talks about his daily schedule which (to my mind) seems to be quite rigid. The couple leave B. an hour to play by herself after she wakes up around 7 a.m. During the day they manage to feed her precisely every 4 h and give her a bottle at 9.30 p.m. in order to get enough sleep. The baby never waves to his wife (see interview above). B. is well accepted by everybody. After Christmas he plans to inform the other parents of his baby group about the trisomy of B.

"If anyone has problems with this: that is his problem and not ours". The only problem they have had was with friends from the medical profession who could not understand that they decided against an interruption of pregnancy. "That's it: they are not our friends anymore… either they accept B. as she is or it means the end of our friendship…" He then talks in detail how they decided to continue pregnancy. The most terrible situation was waiting for the definitive results of the amniocentesis because they feared it could be a trisomy 18. They were so happy to hear that it was "only trisomy 21" and that no severe comorbidity had been discovered. If B. had had severe health problems they would have interrupted the pregnancy. "So your motivation was on medical and not religious grounds…?" Mr. E. nods in approval. He then reports how they looked for a name for B. They chose one which is easy to write because they wanted her to be capable to write her own name later. We have the "goal for our daughter to be able to live on her own by the age of 20 – we don't expect her to be a student or anything like that – but just to live on her own perhaps within a group of handicapped people…" I finally ask him directly if there are moments where he regrets his decision. "Well sometimes I think – did this have to happen? – but this lasts only a few moments. Last week, e.g., she was teething and was clinging to me for 3 whole days – I could not even eat breakfast. This was terrible… But afterwards I thought well she was in pain and then everything worked out again…" He then talks about his son from his first marriage whom he had to give up when his son was three because of his divorce. "With B. this will not happen to me again – I will see her development without any interruption … and this is wonderful…". He has not told his son, who lives in a city quite far away about the birth of B. "He is just starting school and has some problems – I did not want to burden him – I will tell him later on. I only have contact to him by phone, and never see him face-to-face …"

At the end of the interview we talk about the study. We should inform his wife – she will then tell him everything, he says.

Expert validation:

The group members first verbalize their admiration for the vitality of this couple and their capability to cope with difficult situations. They seem to have large and stable resources (see e.g. the fight of Mrs. E. to get her baby out of the children's clinic). The role reversal seems to have many practical advantages (economic etc., health problems of the husband) although some problems seem observable (e.g. that Mr. E. seems to be absorbed by his concern not to be fed "well enough", which we know to be a common concern of many men after the birth of a child). The feeling of being happy might at the same time serve defence purposes (e.g. against the painful feeling of not having a 'normal child' or an avoidance of mourning) but at the same time it seems very functional in order to cope well with the situation. The couple seem capable of enjoying the fact that they have a child together (after two stillborn babies) and to experience this as happiness instead of feeling resentment at not having a 'normal child'. We also discuss the fact that Mrs. E. had been a rejected child given away to her grandmother shortly after her birth. From this angle the experts estimate that loving a child could also have a deep reparative psychological function for her. With pride and amusement she talks about the fact that her parents usually only come to see her after a visit to their hairdresser in town. "Now they come so often that I doubt that the hairdresser has any hair left to cut at all… they just want to see B...."

One "finding" of these interviews was that the decision for or against a handicapped child (e.g. suffering from trisomy 21) seems to be highly individual, depending on biographical factors of the parents as well as aspects of their current life situation. These individual factors seem to be dominant compared e.g. with questions how much support for the handicapped child can be expected from society. There appears to be a connection between the capability to 'love a handicapped child' and handling of personal ambivalences, psychological resources and coping strategies, as well as with stable and relatively mature and stable defensive strategies. The parents' resources to find creative and original solutions for a "good enough" everyday life with a handicapped child seem to be essential.

Interview 6 (substudy B): "Why should you risk such a burden.....PND turning death to life..."

The following two interviews were chosen to shed light on 'positive' possibilities implementing PND in families with a severe genetic problem.

A psychoanalytical colleague answered our letter requesting cooperation in substudy B. He offered to ask a former psychoanalysis patient, Mrs. F., who had undergone PND in the 1990's due to a genetic problem (haemophilia) in her family. I called up Mrs. F. who agreed to be interviewed.

Summary first interview:

Mrs. F. waits in front of my University office for 10 min before the interview appointment; she is a pleasant woman, about 50 years old with short grey hair and expressive brown eyes. On our way to my room we start a pleasant and informal conversation about the current situation in German universities etc. She works with drug-addicted patients as a social worker.

I ask her to tell me about her experiences with PND. She starts to tell her story in a very coherent and impressively communicative way: it proves unnecessary to ask her any of the semi-standardized questions:

She became pregnant for the first time at the age of 19. She decided in favour of an abortion "because I would not have been able to raise a child at that time." At 23 she again became pregnant from her later husband and again decided to abort. "I was in such an instable psychological state at that time – I would not have been capable to function as a mother, nor would my husband. He also agreed to the abortion... I was not capable of really feeling close to someone else. I longed for proximity but always destroyed it myself e.g. by having another lover. I was in such an instable psychological state. My husband X. then got a job in South Africa for 2 years. Paradoxically X. felt much closer to me while living abroad and wanted to have a more stable relationship. I finally decided to visit him there...This decision was in fact quite good for me because I had to face my definitive relationship with X. officially. At that time I developed phobic symptoms – first concerning mice and other small animals, afterwards I was afraid of dogs. Finally, before leaving for South Africa I wanted to sign some cheques for the trip and suddenly started to tremble so intensively that I could hardly write... This symptom was also a problem in South Africa. X. observed it and told me that I needed professional help. When we came back to Germany I looked for psychoanalytic treatment.

The phobic symptoms disappeared very soon. Much more important for me was that I 'found myself' during psychoanalysis: I developed a stable basis within myself. When I was 32 I finally discovered an intensive wish to have a child. I had never thought of it before seriously..." "Was this due to the genetic problem in your family?" I dare to ask, because Mrs. B. had not mentioned this problem although she had talked about it during our first telephone call.

After a short pause she starts to tell the story of her "heavy, burdened childhood": Her parents, relatively poor people without much education, got married when her mother was 19, her father 21-years-old. Their first son suffered from haemophilia. The parents did not know anything about this illness. The family doctor sent them to the University Clinic when U. was 1 ½ years old. "It probably was the way my parents were handling the illness which was so terrible: they could not help my brother because they just did not have adequate knowledge. They both often denied the illness. And then terrible things happened". The doctors fixated his one leg. Once U. was chewing gum and produced a clot of blood beneath his tongue. The doctors feared that this clot of blood could wander into the brain and kill my brother. Therefore he was shut into a room of the clinic fixed in a strange position to keep his head straight. Not until my brother was 29 did my mother discover a specialist on haemophilia in another big German city. My brother finally got adequate medical treatment. He is still a very difficult person, now also suffering from AIDS (having become infected by blood transfusions). He also was very envious of me and often tortured me as his 6 year younger sister.

She then talks again about psychoanalysis. "It has been such an important and good experience for me. I really seemed to discover myself in the depths of my psyche. I had lived behind a kind of white screen for years not really living in the Here and Now of the present. Losing this basic feeling of unreality was the most important development during psychoanalysis and enabled me to finally dare to become pregnant." Only after her wish to become pregnant became real did she undertake genetic testing and discovered that she indeed carries a genetically deformed X chromosome. She became pregnant very soon and underwent 'Chorionic villus sampling (CVS)'. Something went wrong during this medical procedure: She suffered serious bleeding which finally became so dangerous that she had to take medication to prevent labour. In the second interview she talks about her panic attack CVS because she felt helpless and at the mercy of the doctors. She felt the same when under medication for preventing premature childbirth. She found this situation unbearable. Therefore she decided to tell the doctors "to cease medication after some hours and to leave it to nature…" It was very important for her that the nursing staff supported her decision. Without medication labour started. She was narcotized and delivered of a genetically normal boy in the 20th week. After this procedure she wanted to leave the clinic at once: "I did not want my baby to be associated with illness…," she says. She was very sad about the loss of her son.

She felt well treated and supported by the medical staff. The doctors told her to wait 6 months before trying to become pregnant again. She did – and underwent an amniocentesis this time. She got a positive result." I had decided in advance that in this case I will interrupt the pregnancy. It was not such a painful experience as the first time". She then described that a cousin of hers had hurt her with serious reproaches: "How can you select your baby like in a supermarket: This is one I want, and this one is not…… Without analysis I would not have been able to endure such reproaches…" I, as interviewer, support her while she expresses her wish of not wanting her own child to suffer so much as her brother had to during his whole life. "My brother was also very hurt by my decision". It was only in analysis that I comprehended that for him my decision meant: "You don't want a person like me to live – you kill him off!" "Thanks to analysis I was finally able to talk openly with him and try to explain that my decision was not directed against him as a person but against the illness…"

It was very important for her that her husband shared and supported her decision. "I just did not want to continue the darkness of my family. Perhaps you could do better with a handicapped child than my parents did – but I was not sure if I personally could do a good job.…" Her mother had warned her not to become pregnant: "Why should you risk such a burden.....!" she said – pregnancy can mean a catastrophe. Nevertheless she became pregnant again and gave birth to a normal daughter: " It was such great joy and happiness – my daughter is 15 now – and has changed my whole life…" She still seems very proud that she had a "very normal pregnancy and an easy birth". "After 3 h we could leave the clinic and cook spaghetti at home…" she reports with a bright and happy smile.

She sometimes feels sad because she has only one child. "I finished my psychoanalysis because we went abroad again – this would have been an ideal situation to have another child but I did not dare to go through 'the same procedure' again without my analyst. Without the possibility to work through all my feelings and conflicts concerning pregnancy and child abortion I felt too vulnerable to go through it all again… I am so grateful to have one healthy child. My husband feels the same: he was also against a second pregnancy…"

We agree to have a second interview 3 days later.

Supervision

We are impressed by the coherency and the self-reflectiveness of Mrs. F. She talks about deep insights in a very personal language and way of expression; impressive is also Mrs. F.'s capacity for creativity and strength to find a 'good life' in spite of severe traumatisation during childhood. We suppose that her yearlong feeling of 'living behind a screen' could be due to her severe traumatisation and indicate dissociative states (as we often find them with severely traumatized patients).

We wonder if the development of phobic symptoms could unconsciously have been connected to her two abortions. If possible we would like to ask Mrs. F. in the second interview if she had thought about this possibility in psychoanalysis.

Second interview

The good emotional rapport between Mrs. F. and myself is resumed at once. I ask her if the first interview has evoked too many painful memories in her. She denies this and tells me that it was important opportunity for her to talk about all these experiences again. She then surprises me with her remark: "Maybe I am not the right person to talk with for your study. I had already made my decision before I was confronted with the positive findings. Maybe women who never had to think of the possibility of having to decide on life or death of their unborn child experience the ethical dilemma in a more intensive way then I did…" I am impressed by Mrs. F.'s empathy and identification with the aims of our study.

She now talks about her sister. After an abortion she did not become pregnant for 5 years. She separated from her husband and fell in love with another man: she then became pregnant (at 37 and then again at 41). "She has no genetic deficiency – but she was of an age where many women choose amniocentesis. My sister refused this – she took the risk of giving birth to a handicapped child…" The sister has a better relationship with her brother then Mrs. F. "She was the third one: my brother did not feel so envious towards her as with me…"

I now dare to ask if she felt guilty because of the early abortions and if – in psychoanalysis – they had found out that abortion was involved with the development of the phobic symptoms. She denies my hypothesis: "No – I felt no guilt because of the abortion. I was so much in love at 19 – we felt like little children, having sex without any precaution for a whole year. We never thought about possible pregnancy – which all sounds very childish to me now. My ex-boyfriend thinks similarly: I am still in contact with him. He also thinks that abortion was the best step to take at that time…

Nor do I feel guilt about the second abortion: my husband and I agree that we would not have been capable parents for a child…" We talk about the 'Zeitgeist' during these years: she was influenced by the feminist movement with its slogan: "Mein Bauch gehört mir…" (My belly belongs to me). Abortion was not forbidden or taboo in these groups at this time. In psychoanalysis they understood finally that the phobic symptoms of not being able to sign a cheque was unconsciously connected to the taboo to 'demonstrate' that she had become an independent and sexually mature woman – not a child anymore. Signing had the meaning of publicly demonstrating this truth!

She remembers that her mother had strongly tabooed sexuality. When she used some lipstick in her early adolescence her mother used to say: "Are these early signs of a prostitute?" In her early childhood she always felt more attracted to her father than her mother. "My father used to smell so beautifully – I liked to sit on his lap… Even now I feel closer to him and feel a need to be apart from my mother. I hate her constant complaining and even blaming others for her own terrible fate… I never want to be like that. When something bad happens to me, I always try to do something actively against it. E.g. when I got the news that I am a 'transmitter of haemophilia' I did not complain or blame my parents for this fate: I just thought I will have to make the best of it…" 'Turning passivity into activeness..' became one of her major character traits. "In analysis we understood that the wish to actively do something against bad things is sometimes exaggerated. I can hardly bear to just wait and suffer even for a short time without going into action right away..." "Did your father have a similar outlook?" "No, he was a rather anxious person. He worked in his father's little firm, a terrible patriarch who decided on the life of my father and humiliated him publicly again and again... He died at the age of 58 from his third heart attack.... Therefore I think I developed the active way of problem solving myself... I am still very proud that I left home at only 17 – to go to a school in another town which I had chosen all by myself, without my parents even knowing about my application..." We then talk about the sunny and shadowy sides of this early vital and extreme flight into autonomy during adolescence. Mrs. F. describes her insights during analysis: she also felt very much alone during these years "without a core making me feel at home within myself...." She already developed first phobic symptoms during these years.

She again describes her feeling as living behind a screen for years during adolescence and late adolescence. As I ask her if this could be a defence against her fear of being flooded by sudden terrible traumatic events again as she had repeatedly experienced during her childhood, she remembers that her mother used to reproach her: "You were a child of very little empathy. When your brother had to be taken to the hospital in the middle of the night – and the ambulance came to pick him up with sirens blaring and blue lights flashing – you did not even used to wake up. You slept like a bear...." She agrees that these reactions might have been a psychological attempt to build up a protective shield against sudden traumatic events. "I think that my symptom of living behind a screen or looking at things only through a narrow tunnel – could have been the continuation of this coping and defence strategy of my childhood..."

At the end of the interview she asks me if we would also be interested in interviewing a young woman from another genetically burdened family who has just given birth to a normal child after having undergone PND. She wants to ask this woman if she would be willing to be interviewed.

Saying goodbye, she again tells me how grateful she feels that psychoanalysis has enabled her to give birth to a healthy child "and to leave the darkness of my childhood behind me..."

Summary of the experts' validation

All the experts agree that Mrs. F. has achieved an impressive way of dealing and coping with the severe traumatisation of her childhood. They support the hypothesis that for Mrs. F. the main achievement of psychoanalysis had been to develop her own stable core personality and to overcome the feeling of "not really living on this Earth". The feeling of living behind a screen is probably connected to dissociated states documented in clinical work with severely traumatized patients. Thus, probably the first two abortions were not experienced as 'reality' at the time, perhaps one reason why Mrs. F. did not develop any guilt feelings.

Mrs. F. gives an impressive example for the necessity of professional help in a situation where decisions have to be made for or against the life of an unborn child. 'Playing God' probably overtaxes the psychological capabilities of many women. Mrs. F. can describe how painful her guilt feelings were when deciding to abort the handicapped embryo – a human being like her brother. Maybe this is one reason why she likes working with drug addicts: "It is so touching for me to see that some of the alcoholic patients are able to stop drinking and become healthy again..." "You can stop drinking alcohol – but you can not stop being a haemophiliac..."

Mrs. F. also described how helpful and important the medical staff and the support of her husband and the analyst had been for her. "Without analysis I would have never dared to become pregnant...." She needed professional therapeutic support in order to be able to separate from the darkness of her childhood and decide for a healthy child, "on the sunny side of life..."

1.6 Some Concluding Psychoanalytic Remarks on Clinical Findings of EDIG by Interviewing Couples After PND

Using these selected interviews with three women/couples after PND I have illustrated how the personal experiences and coping with the ethical dilemmas can be both complex and highly idiosyncratic. By listening so intensively to women/couples who had undergone PND and communicating with their experiences I also hoped to transmit some insights on possible short-term and long-term effects of late pregnancy interruption or a decision to give birth to a handicapped child. I attempted to summarize the interviews relatively close to the verbalization of the interviewee and, taking into account the methodological considerations mentioned above, have been careful with psychoanalytical interpretations. We used expert validation mainly as a possibility to try to comprehend the conscious and unconscious communication of the interviewee as cautiously and correctly as possible.

In spite of the highly individual dimension of experiencing ethical dilemmas due to PND we would like to conclude this chapter with some general psychoanalytical remarks concerning our clinical findings. They are based on the exchange with 45 women/couples after PND who have been interviewed in the Frankfurt/Mainz group to date. The second source consists of interviews with 16 psychoanalytical colleagues concerning their psychoanalytical insights gained in long-term psychoanalyses with women after interruptions of pregnancy (mostly performed by me). Of course these considerations are preliminary and primarily aim at formulating some hypotheses on risk and protective factors for women/couples undergoing PND that require further investigation in the future.

For all the women/couples the unexpected confrontation with a positive finding of PND, the necessity of deciding on life or death of an unborn child, and particularly the experiences of late interruption of pregnancy were extremely burdened. For some of them they were even of traumatic[6] quality. As already mentioned, according to Bohleber (2000, p.798) experiencing trauma can be characterized as having to cope with a situation, which has the quality of "too much" (in respect of the so-called oeconomical as well as the object relational model of contemporary psychoanalysis). Cooper (1986) defined psychic trauma as an event, in which the capability of the ego to guarantee a minimal feeling of safety and integrity is suddenly overwhelmed, evoking overflowing anxieties and a state of mind of extreme helplessness. Such a traumatic event leads to long lasting changes of the psychic organization (see also Fischer and Riedesser 1998). From an object relational perspective the basic trust in a helpful, 'containing' and empathetic inner object is destroyed because this 'good inner object' was not capable of protecting the self in the traumatic situation. In all the interviews summarized in this chapter, the overwhelming, and in some cases traumatic quality in connection to PND was obvious. We would expect that for some of the women/couples these experiences indeed

---

[6] I cannot go into the current debate on trauma here. This term is often used in an imprecise, broad way und thus in danger to lose its explanatory power (see e.g. Leuzinger-Bohleber et al. 2008a, 2008b).

have the long-lasting effects of the trauma described above. But why do our empirical findings replicate the findings of other studies that for the majority of the women/couples the experiences seem to have the quality of a relatively short crisis, which can be overcome within a relatively short period of time whereas for others this is not the case? How can we differentiate between these two groups?

Most psychoanalysts who have been interviewed agree on the following psychoanalytic considerations:

> A woman (or couple) going through an extremely burdening (or even traumatic) situation (such as being confronted with an unexpected, shocking diagnosis e.g. "the child has no face...", or having to give birth to a dead child) has to mobilize extreme forms of coping – and defence strategies in order to 'survive' the acute (traumatic) situation. For psychic reasons the complexity of the situation has to be reduced radically in order to enable a decision (e.g. concerning the life or death of the unborn child within a few hours or days or to go through the birth of a dead child) and to act in this extreme situation. If these experiences do have a traumatic quality, then often a specific coping mechanism, so called "dissociation", can be observed: The self dissociates from its emotions, fantasies, thoughts in an extreme way. It 'flees' into a different state of mind, which (on the surface) has nothing to do with the overwhelming emotions and fantasies evoked in the traumatic situation. The individual – at first sight – can function surprisingly well, is e.g. able to work and cope with everyday situations shortly after the traumatic event. But at the same time the individual has lost the inner connection to one's self, its own emotions and thoughts, to the object (e.g. the partner) and to the "real quality of life". This state of dissociation is often not recognized by the person her- or himself and not connected to the traumatic situation, e.g. with the loss of the baby etc. As we know from long-term psychoanalysis such dissociative states may sometimes endure for years and – unconsciously – determine the psychic reality of the individuals (see e.g. Mrs. F.). The severely traumatized persons hardly find their way back to 'normal life' again, and are not really fully living in the present again. They have completely lost touch with the ground under their feet. They generally don't feel any rapport to other persons anymore and have lost the basic feeling of being the active center, the motor of their own lives (see also Leuzinger-Bohleber et al. 2008a, 2008b). For some of the women/couples investigated in substudy B this seemed to have been the case: Even after years after the interruption of pregnancy they have not regained psychic normality. Others seem to overcome the crisis and even psychic states of dissociations within a relatively short period of time. While listening to them you get the impression that they were able to finally integrate the extremely burdened experiences into their selves and identities often with the help of 'empathetic others', supporting persons in their private lives or in professional contexts. Why are experiences after PND for some women/couples traumatic and others just severely burdened?

In order to clarify these questions let us take into account some further considerations based on the classical structural model of psychoanalysis and some newer theoretical approaches in psychoanalysis (see Leuzinger-Bohleber 2008c). According to clinical observations the extreme (traumatic) situation of suddenly being confronted with life and death of one's own child either reactivates[7] an

---

[7] In the situation of PND – in which you have to chose between life and death of the unborn child mostly within a very short period of time – you have hardly any open space (or what in psychoanalysis is called "intermediate space") in order to experience and consider the many facettes of your decision. The structure of this situation, having to clearly decide between "yes" or "no", shows analogies to a preambivalent state of the mind, the so called "paranoid-schizoid position". Therefore a reactivation of this archaic state of the mind is highly probably in this situation.

"archaic"[8] state of psychic functioning or leads to extreme regression into an archaic state of psychic functioning: As Freud (1926) has already described, the confrontation with one's own death or the death of a close and beloved person (particularly one's own child) is an overtaxing psychic situation and absorbs all the psychic energy at once. Death anxiety is the most extreme form of anxiety, which mobilizes primitive coping and defence strategies. We think of 'primitive' mechanisms like denial, splitting, projections and projective identifications etc. (see Moser 2009). This archaic state of psychic functioning is dominated by a pre-ambivalent state of the mind, which Kleinian psychoanalysts called the paranoid-schizoid position. This state of mind is characterized by an extreme psychic split between 'bad' and 'good', 'black' and 'white', victims and persecutors. Connected with this state of mind is the reactivation of an archaic world of unconscious fantasies on murderers and innocent victims, witches and saints, devils and angels etc. The Medea fantasy, mentioned in the introduction of our publication (Leuzinger-Bohleber et al. 2008a, 2008b) seems to be one example of such an ubiquitous female (body) fantasy. Many women, deep in their unconscious world, are convinced of having the potential to become the murderer of their own children, particularly in a situation of extreme helplessness, narcissistic vulnerability and overtaxing loneliness. Unconsciously they thus fear revenge and suffer from unbearable guilt feelings. Also (archaic) oral fantasies are often evoked, e.g. the fantasy of an oral conception of the baby, of having poisoned the fetus e.g. by smoking, drinking alcohol or having taken dangerous medication. The 'fact' that the fetus had to be 'eliminated' in order not to threaten the life of the mother, may also evoke anal fantasies (see e.g. Mrs. and Mr. C. both seemed to be psychically absorbed with fantasies about the "dirty, terrifying fetus" on the photo in the enclosed envelope, which had to be expelled. It seemed like a kind of reaction formation that they so urgently asked for a friendly, clean room with flowers etc.). In other interviews and therapies we could observe that oedipal fantasies had been reactivated in the context of PND. Five psychoanalysts reported that the disability of the fetus – unconsciously – was experienced as a punishment for infantile oedipal wishes. The deformation of the fetus and the interruption of pregnancy were – unconsciously – thus seen as revenge or a punishment for such forbidden wishes.

The regression onto this archaic level of psychic functioning with a primitive, pre-ambivalent logic of good and bad, right and wrong – as well as the reactivation of the just mentioned archaic unconscious fantasies – may be some of the psychic sources for the unbearable quality of shame and guilt feelings. As illustrated in many interviews and reports from psychoanalyses, many of the patients who had interrupted

---

[8] "Archaic" means a psychic functioning which seems typical of an early developmental stage. It is often used in psychoanalysis as an opposite term to a "mature" level of psychic functioning. Due to the limited mental processes of such early developmental stages, we could metaphorically talk of a "primitive attempt" to find orientation by clearly discriminating between "good" and "bad", "black" and "white". Psychoanalysts speak of a preambivalent way of psychic functioning.

their pregnancy seemed unconsciously convinced of being a child murderer, always magically expecting revenge and punishment. To mention just one exemplary case:

> 50-year-old Mrs. H. was seeking therapy, because she suffered from heavy depressions. At the age of 44 she underwent a feticide, because she was 6 months pregnant with a trisomy 21 child. The ultrasonic test additionally showed further abnormalities (deformation of the heart, reduced growth, etc.). The doctors advised her to interrupt pregnancy, as there was a high probability the child would not be able to live. Her husband and the whole family supported her in this decision. Nevertheless, during psychoanalytical treatment it became obvious that Mrs. H. suffered from severe unconscious feelings of guilt and that she felt like the 'murderer' of her child, which is one of the major reasons for her depression. She was haunted by obsessional fantasies which dealt with the memory of the deadly needle during the feticide (she and her husband were watching, and they could see how long it took). She had to carry the dead child in her womb for 4 days until childbirth could be initialised, another source of archaic unconscious oral fantasies: Who is killing whom – the mother the baby or the other way around? Only when these fantasies could be worked through in the therapeutic relationship the depressions more or less disappeared.

From a psychoanalytical point of view[9] a confrontation with one's own unconscious archaic world of a murderous self and other[10] can hardly be prevented going through a late interruption of pregnancy in which a 'murdered child' is indeed part of 'reality'. This outside reality is often then confounded with the 'archaic inner reality' into which the traumatized person has regressed. To get in touch with this state of mind and the archaic (of course unrealistic!) quality of the fantasies and its characteristic psychic functioning is a presupposition for overcoming regression both after and in acute traumatisation. This means returning onto a more mature level of psychic functioning where you can 'rediscover' and experience the complexity and the ambivalences, which are always connected to PND. This is indispensable in order to overcome the trauma or the extreme crisis and to regain psychological health. The empirical finding (mentioned by Fischmann in this volume), that women/couples who had interrupted pregnancy due to a positive finding of PND and who expressed their impression that "there is no right decision in such a situation", might be an indicator for these psychic processes.

---

[9] Of course we do not have the space in this volume to elaborate on the basic finding of psychoanalysis that all infantile fantasies are kept in the unconscious. In normal, everyday situations they do not disturb the psychic functioning of a "healthy adult person" because they are psychically well integrated and kept in the unconscious by functional defence processes. But in extreme (traumatic) situations, with some structural analogies to the original infantile experiences (e.g. to become pregnant bears some similarities with being fed by another, existentially important person as a baby), these unconscious fantasies can be reactivated and then determine the current psychic realities. We think that the traumatic situation after a positive finding in PND is structurally predetermined to reactivate early experiences on life and death which have been kept in an archaic dimension of one's unconscious. By the way: this psychoanalytic thesis is given new support by contemporary neuroscience research which also postulates that "the brain does not forget anything…" (see Leuzinger-Bohleber et al. 2008a, 2008b).

[10] Working psychoanalytically with women in such a traumatic situation often reveals, that unbearable own murderous impulses are projected onto the partner or the medical doctor, a psychic possibility to "morally fight" them as attitudes of the partner and the doctors vehemently.

Psychoanalysts know that in extreme (traumatic) situations, where an individual is confronted with life and death, the personal inner psychic resources mostly prove to be insufficient in order to cope in a mature and healthy way with the overtaxing situation if one has to cope with the situation totally alone, without support by emotionally close persons or a supporting social environment. Therefore, all the women/couples undergoing late interruptions of pregnancy due to PND absolutely need a counterpart to their reactivated archaic inner world in their outside reality, in a loving partnership, the family, friends, but also in the professional medical care during PND in order to finally overcome the regression and to regain a more mature level of psychic functioning after a relatively short crisis. If such support is not forthcoming the crisis might culminate in a dramatic process in which the individual is in desperate need of professional help in order to cope. The therapist proves to be a "good real object" that helps the patient to find her way back to reality, out of the nightmare of inner persecutions, archaic guilt feelings, shame and despair. Only after such therapeutic working through is a process of mourning – and psychic healing – possible.

For overcoming the acute crisis not only the support from the outside reality but also a stable inner object world of the individual itself might be protective factors which enable the majority of the women/couples to regain their psychic equilibrium, or what Kleinian analysts call the "depressive position", a capability to cope with mature ambivalences and complexities again. If "good inner objects" can be mobilized in the acute crisis and can be supported by the experience with "good objects" in the outside reality, feelings of loss, guilt and shame can be experienced in a more mature way and lose their persecutory quality.

The patients, who – like Mrs. H. – became severely depressed years after the interruption of pregnancy, often do not have such protective factors. Often they have not gone through stable "good enough" early relationships and thus not become able to develop stable inner objects, good narcissistic self-regulation, a secure attachment pattern as well as a functional capability to symbolize and mentalize. Often a relatively archaic level of psychic functioning is dominating: guilt feelings, aggressions and coping and defence mechanisms thus often have an archaic quality. They often have gone through intensive trauma during early childhood and/or adolescence. Particularly traumatic losses of close relatives or even their own child often prove to be risk factors. These women/couples should get professional help overcoming the acute crises in connecting with a positive finding after PND.

Protective and risk factors, seen from a psychoanalytical standpoint, are thus due to idiosyncratic, biographical characteristics of one's own inner psychic world, which are not easily diagnosed because they are not directly observable. Nevertheless experienced clinicians learn how to perceive (and afterwards to test) some indicators for important features in the psychic reality. These indicators for protective and risk factors for women/couples undergoing PND can be transmitted to nonpsychoanalytic persons, e.g. medical doctors and their staff.

These indicators can best be directly observed in the interaction between women/couples and their medical staff before, during and after PND itself. In the following table we summarize some of these indicators for protective and risk factors: If professionals bear these protective/risk factors in mind, they might be able to observe their manifestations in direct interactions with the woman/couple while confronting them with problematic findings of PND. To give just a few examples: If a woman/couple can not show more or less adequate emotions and thoughts (indicator for a dismissed attachment pattern) or is extremely overflowed by affects and fantasies (indicator for a preoccupied or desorganized attachment pattern and a lack of mentalization) the professional person should take a closer look and actively ask about additional risk factors such as traumata in the biographies of the woman/couple, former losses etc. Another indicator for a vulnerable psychic state of the woman/couple could be if she (or he), after getting the problematic news, at once looks for someone else who could – undoubtedly – be guilty or responsible for the findings. This could be an indicator that the person is not capable of enduring ambivalent feelings and 'mature, adequate' guilt feelings for taking the responsibility for one's own situation. It could also be an indicator for a dominance of primitive defence mechanisms as denials, splittings and projections in her/his psychic reality in the sense of the above mentioned paranoid-schizoid position.

Protective and risk factors, seen from a psychoanalytic standpoint, are thus due to idiosyncratic, biographical characteristics of one's own inner psychic world. In the following table we try to summarize some of these indicators Table 1.2.

1.7 Suggestions for Training of Medical Staff

Of course doctors and medical staff should be very careful with these hypotheses and try to elaborate them in more detail together with the specific woman/couple. But the findings of the interviews in substudy A and B very clearly show that an empathic and professional dialogue with the medical doctor who confronts the women/couple with the problematic finding is essential for the short-term as well as the long-term coping with a problematic finding of PND. In the best case the knowledge might help professionals to be 'good real objects' to women/couples in the traumatic situation after positive findings in PND. This could help them to deal with the above mentioned regressive processes into their own inner archaic psychic world and going through the inevitable crisis in a more or less productive way, asking for support from family, friends and professionals and thus increasing the probability to regain a mature psychic level of functioning soon. Therefore, special awareness training for doctors and medical staff in the field of PND seems helpful and even necessary.

Of course more research is needed in order to systematically test these clinical findings. All of the above mentioned indicators are mere indicators, not yet 'objective, absolutely reliable findings'. Nevertheless, we think that EDIG can be considered an important step in order to develop training programmes for professionals with the aim to improve concrete counselling practice in the field of PND. One of the aims of such training could also be that the medical staff is capable to diagnose those women/couple who need professional help in order to overcome the acute crisis due to PND.

Table 1.2 Protective and risk factors

Protective factors			Risk factors		
Inner world	Indicators	Indicators during PND	Inner world	Indicators	Indicators during PND
Good inner objects	Stable relationships with partner, friends, family	Trustful, open contact with medical doctors, staff etc.	Fragile, unstable inner objects	Rarely good relationships to partner, friends, family, socially isolated	Estranged relationship with doctors and medical staff (no basic trust, difficult communication)
No severe traumatic experiences in childhood and adolescence	Integrated personality (good integration of emotions, fantasies, thoughts, no dissociative state, can communicate own anxieties, concerns, meets persons trustfully	Can communicate with medical doctors and staff in an uncomplicated way, talks about feelings, anxieties etc., shows curiosity, openness for information etc.	Traumatic experiences in childhood and adolescence	Traumatic experiences influence communicative style, no basic trust towards partner, friends, professionals etc.	Not integrated emotions, signs of dissociations, mistrustful attitude towards staff and medical doctors
No former loss of child or close person		Loss is not spontaneous topic in contact to medical doctor or staff. When asked directly the person affirms the lack of serious losses	Former loss of child or close person		Person may seem depressive, anxious, talks about former losses, when asked directly the person affirms serious losses in her/his biography
Secure attachment	Individual has access to broad range of feelings, ambivalences etc.	Individual can show broad range of feelings (e.g. anxieties, despair etc.) when the doctor confronts it with the problematic findings,	Insecure attachment	Individual has either to deny negative emotions (dismissed attachment pattern) or is overflowed by them (preoccupied or disorganized attachment pattern)	Individual either does not show any emotions while confronted with the problematic findings or is overflowed by panic and despair in an extreme way

1 Ethical Dilemmas Due to Prenatal and Genetic Diagnostics

Mature level of psychic functioning e.g. capable to experience ambivalences, mourning etc. ("Depressive position")	Individual can perceive and express ambivalent feelings and thoughts	Individual can perceive and express ambivalent feelings and thoughts in talking to medical doctors or staff	Archaic level of psychic functioning, preambivalent state of the mind ("Paranoid schizoid position")	Individual splits in extreme ways between "good" and "bad", "right" and "wrong" aspects of PND, decisions etc.	Individual splits in extreme ways between "good" and "bad", "right" and "wrong" aspects of PND, decisions etc. while talking to doctors or medical staff, looks for another person who "undoubtedly" is responsible for the finding
Mature coping with guilt feelings[a]	The self can stand to become guilty without a basic feeling of being destroyed in his self concept and esteem fundamentally	Individual can talk about guilt, shame etc. in a "mature", adequate way	Archaic guilt feelings	The guilt feelings have an archaic, unbearable quality and therefore have to be split of, projected, denied etc.	Individual does not seem to have guilt feelings, others are blamed for the situation, one's decision etc. individuals often show strange psychosomatic reactions instead of direct emotions
Mature quality of aggression	Individual can perceive and accept own aggressive impulses because they are not mainly associated with destruction	Individual can show or talk about aggressive, non destructive impulses	Archaic quality of aggression	Individual denies, splits of, projects own aggressive impulses, fears revenge, destructiveness of others etc.	Individual can not show directly aggressive impulses or is overflowed in an uncontrolled way by them, often individual feels to be the passive victim - others are the persecutors

(continued)

Table 1.2 (continued)

Protective factors			Risk factors		
Inner world	Indicators	Indicators during PND	Inner world	Indicators	Indicators during PND
Stable narcissistic self regulation[b]	Individual has stable narcissistic self-regulation in private relationships, job etc., well developed autonomy	Individual shows a socially adequate behaviour towards doctors/medical staff (can accept medical authority without too much submission), seems to be able to use information in an autonomous way	Fragile narcissistic self regulation	Individual needs strong narcissistic support by other persons, is narcissistically vulnerable	Individual seems to be in a constant vulnerable stage, often feels insulted (by medical doctors, staff), does not show much autonomous thinking, actions etc.
Dominance of mature defence mechanisms	Individual can use sublimations, rationalizations, intellectualizations	In contact with medical staff person is able to show intellectual interest without losing emotional contact	Dominance of primitive (archaic) defence mechanisms	Psychic life seems to be dominated by denial, splitting, projections, reversal in the opposite mode etc.	In contact with medical doctors the individual seems to deny important information, splits between "good and bad", tries to project negative feelings onto others
Mature coping strategies	Individual has a range of mature coping strategies dealing with difficult situations (in professional and private situations)	Individual can take up advices, suggestions by the medical staff and completes them with own suggestions, ideas etc.	Lack of, or primitive coping strategies	Individual can hardly solve difficult situations alone, is extremes dependent on the advices of others	Individual shows extreme helplessness and "infantile" ideas on how to cope with the difficult situation of PND, can hardly ask relevant questions etc.

| Sensitive for cultural and ethical factors of PND | Individual is capable to reflect on cultural and ethical issues of PND | Individual takes up cultural and ethical questions in the consultation with medical doctor and staff | Not sensitive to cultural and ethical factors of PND | Individual seems to live in its own personal world in an extreme way | Individual mentions strange connections between PND and own situation, is not capable to reflect on cultural and ethical issues |

[a]Examples of a mature level of psychic functioning can be found e.g. in the interviews with Mrs. F.

[b]This means a stable psychic possibility to regulate one's own self-esteem, in psychoanalytic terms a stable regulation of one's narcissistic needs. These processes are very important for psychic health. A stable narcissistic regulation of the self is a presupposition for mature love. To say it in a simplified manner: You can only love another person in a "mature way" if you are able to love yourself in the same "mature way". The psychic level of the self and the inner objects is mature. So called self-psychological authors in psychoanalysis have studied these aspects of psychic development particularly intensively. They have elaborated that a stable mature way of narcissistic self regulation can only develop if the primary objects empathise with the narcissistic needs of the baby and the small child and help them to cope with frustrations in order to psychically integrate archaic fantasies of the grandiose self and the omnipotent parent imago.

1.8 Containing Functions of Culture and Society for Women/Couples Undergoing PND

As briefly mentioned above: The traumatic quality of the decision on life and death of one's own child often overtaxes women and couples after PND. Helpful 'good objects' in the professional world surrounding them in this situation seem very important. But many of the interviewed women/couples also mentioned how important the public discourse on PND, handicapped children and, more general responsibility for the next generation has been for them. As will be discussed in other chapters of this volume society and culture seem to have a 'containing function' (Bion) for couples in the described traumatic situation which should express empathy and understanding for the extremely difficult situation modern prenatal and genetic diagnostics may lead individuals into, individuals who are members of our society and culture and who have to be supported, and not devalued or even condemned for their decisions. We should also not forget that economic interests might play an increasing role in placing direct or indirect pressure on a couple that decides to give birth to a handicapped child (cf. interview with Mr. E., also e.g. Läzer 2008). As we have discussed in our book (particularly in part IV in Leuzinger-Bohleber et al. 2008b), the decisions evoked by PND for or against giving birth to a severely handicapped child touch dimensions which go beyond the responsibility of the individual couples and have also to be backed up by the societies.

References

Bassin D, Honey M, Mahrer-Kaplan M (1996) Representations of motherhood. Yale University Press, New Haven

Black RB (1989) A 1 and 6 month follow-up of prenatal diagnosis patients who lost pregnancies. Prenat Diagn 9(11):795–804

Bohleber W (2000) Die Entwicklung der Traumatheorie in der Psychoanalyse. Psyche 54:797–839

Bowles Biesecker B, Marteau TM (1999) The future of genetic counselling: an international perspective. Nat Genet 22(2):133–137

Brähler E, Meyer A (eds.) (1991) Psychologische Probleme in der Humangenetik. Jahrbuch der medizinischen Psychologie, 6th edn. Springer, Berlin

Brown J (1989) A piece of my mind. The choice. J Am Med Assoc 262(19):2735

Burnett P, Middleton W, Raphael B, Martinek N (1997) Measuring core bereavement phenomena. Psychol Med 27(1):49–57

Chambers HM, Chan FY (2000) Support for women/families after perinatal death. Cochrane Database Syst Rev. 2000;(2):CD000452. Review

Clarke A (1994) Genetic counselling: Practice and principles (Professional ethics). Routledge, London

Cooper A (1986) Towards a limited definition of psychotrauma. In: Rothstein A (ed.) The Reconstruction of trauma: its significance in clinical work. International Universities Press, Madison, pp 41–56

Elder SH, Laurence KM (1991) The impact of supportive intervention after second trimester termination of pregnancy for fetal abnormality. Prenat Diagn 11(1):47–54

Ethikrat N (2003) Genetic Diagnosis Before and During Pregnancy. Opinion, Berlin

Faimberg H (1987) Das Ineinanderrücken der Generationen. Zur Genealogie gewisser Identifizierungen. Jb Psychoanal 20:114–143
Fischer G, Riedesser P (1998) Lehrbuch der Psychotraumatologie: Mit 20 Tabellen. UTB für Wissenschaft. Reinhardt, München
Freud S (1908) Creative writers and day-dreaming. SE 9:143–153
Freud S (1926) Inhibition, symptom and anxiety. SE 20:87–172
Giovacchini SL (1979) The dilemma of becoming a woman. In: Sugar M (ed.) Female adolescent development. Brunner/Mazel, New York, pp 253–273
Girardon-Petitcolin M (2001) Une figure de la séparation impossible: L'interruption médicale de grossesse. Rev Fr Psychoanal 62:425–436
Green JM, Statham H, Snowdon C (1992) Screening for fetal abnormality: attitudes and experiences. In: Chard T(ed.) Clinics in developmental medicine. Obstetrics in the 1990s current controversies, vol 123/124. MacKeith, London, pp 65–89
Green JM, Snowdon C, Statham H (1993) Pregnant women's attitudes to abortion and prenatal screening. J Reprod Infant Psychol 11(1):31–39
Gregg J (1988) Codes, conscience and conflicts: ethical dilemmas in midwifery practice. Midwives Chron 100(1199):392–395
Hohenstein H (1998) Störfaktoren bei der Verarbeitung von Gefühlen in der Schwangerschaft. Gesellschaftliche und ethische Hintergründe der Fruchtwasserpunktion: Interviews mit Betroffenen und Erörterung ihrer Erfahrungen. Waxmann, Münster
Hunfeld JA (1995) The grief of late pregnancy loss: a four year follow up. PhD thesis, Erasmus University, Rotterdam
Hunfeld JA, Wladimiroff JW, Passchier J, Venema-Van Uden MU, Frets PG, Verhage F (1993) Emotional reactions in women in late pregnancy (24 weeks or longer) following the ultrasound diagnosis of a severe or lethal fetal malformation. Prenat Diagn 13(7):603–612
Iles S, Gath D (1993) Psychiatric outcome of termination of pregnancy for foetal abnormality. Psychol Med 23(2):407–413
Langer M, Reinold E (1993) Somatopsychische Reaktionen auf diagnostische Verfahren während der Schwangerschaft. Int J Prenat Perinat Psychol Med 5:425–431
Lasker JN, Toedter LJ (2000) Predicting outcomes after pregnancy loss: results from studies using the Perinatal Grief Scale. Illn Crisis Loss 8:350–372
Läzer KL (2008) Ethische Dilemmata im Zusammenhang mit genetischer und pränataler Diagnostik aus der Sicht der Betroffenen. Eine inhaltsanalytische Untersuchung. Unveröffentlichte Diplomarbeit, Universität Berlin
Leuzinger-Bohleber M (2000) Psychoanalyse – Erfahrungswissenschaft des Unbewußten. In: Hampe M, Lotter M-S (eds.) Die Erfahrungen, die wir machen, sprechen gegen die Erfahrungen, die wir haben. Über Formen der Erfahrung in den Wissenschaften. Duncker und Humblot, Berlin, pp 145–167
Leuzinger-Bohleber M (2001) The 'Medea fantasy'. an unconscious determinant of psychogenic sterility. Int J Psychoanal 82:323–345
Leuzinger-Bohleber M (2002) A follow-up study-critical inspiration for our clinical practice. In: Leuzinger-Bohleber M, Target M (eds.) Outcomes of psychoanalytic treatment: perspectives for therapists and researchers. Whurr, London, pp 143–177
Leuzinger-Bohleber M (2006) Kriegskindheiten, ihre lebenslangen Folgen – dargestellt an einigen Beispielen aus der DPV Katamnesestudie. In: Radebold H, Hut G, Fooken I (eds.) Kindheiten im Zweiten Weltkrieg. Juventa, Weinheim
Leuzinger-Bohleber M, Stuhr U, Rüger B, Beutel M (2002) Forschen und Heilen in der Psychoanalyse: ergebnisse und Berichte aus Forschung und Praxis, 1st edn. Kohlhammer, Stuttgart
Leuzinger-Bohleber M et al (2008a) Introduction and overview. In: Leuzinger-Bohleber M et al (eds.) The janus face of prenatal diagnostics. Karnac, London, pp 3–43
Leuzinger-Bohleber M et al (eds.) (2008b) The janus face of prenatal diagnostics. Karnac, London
Leuzinger-Bohleber M, Roth M, Buchheim A (2008c) Psychoanalyse – Neurobiologie – Trauma. Schattauer, Stuttgart

Marteau TM, Mansfield CD (1998) The psychological impact of prenatal diagnosis and subsequent decisions. In: O'Brien PMS (ed.) Yearbook of obstetrics and gynaecology. RCOG, Royal College of Physicians (RCP), London, pp 186–193

Marteau T, Richards M (eds.) (1996) The troubled helix: Social and psychological implications of the new human genetics. Cambridge University Press, Cambridge

Marteau TM, Dormandy E, Michie S (2001) A measure of informed choice. Health Expect 4(2):99–108

Moser U (2009) Eine Theorie der Abwehrstrategien. Brandes & Apsel, Frankfurt am Main

Neimeyer RA (1999) Narrative strategies in grief therapy. J Constructivist Psychol 12(1):65–85

Pines D (1993) A women's unconscious use of her body: a psychoanalytical perspective. Virago, London

Prigerson et al (1995) Psychiatry Research. Mark Miller 59(1–2):65–79

Rapp R (1984) XYLO: A true story. In: Arditt R, Duelli-Klein R, Minden S (eds.) Test-tube women: what future for motherhood? Pandora, London

Raz A (2004) Important to test, important to support: attitudes toward disability rights and prenatal diagnosis among leaders of support groups for genetic disorders in Israel. Soc Sci Med 59(9):1857–1866

Ringler M, Langer M (1991) Das Wiener Modell: Ein interdisziplinäres Betreuungskonzept für werdende Eltern bei Diagnose "fetale Missbildung". In: Braehler E, Meyer A (eds.) Psychologische Probleme der Humangenetik. Springer, Berlin, pp 123–138

Salvesen KA, Oyen L, Schmidt N, Malt UF, Eik-Nes SH (1997) Comparison of long term psychological responses of women after pregnancy termination due to fetal anomalies and after perinatal loss. Ultrasound Obstet Gynecol 9:80–85

Sandler J, Sandler A (1983) The 'second censorship', the 'three box model' and some technical implications. Int J Psychoanal 64:413–433

Staham H, Weaver J, Richards M (2001) Why chouse Caesarean section? The Lancet, 357:635

Statham H (2002) Prenatal diagnosis of fetal abnormality: the decision to terminate the pregnancy and the psychological consequences. Fetal Matern Med Rev 13:213–247

Statham H (2003) The parents' reactions to termination of pregnancy for fetal abnormality: From a mother's point of view. In: Abramsky L, Chapple J (eds.) Cheltenham prenatal diagnosis. the human side, 2nd edn. Nelson Thornes, Cheltenham, pp 182–196

Statham H, Solomou W, Chitty L (2000) Prenatal diagnosis of fetal abnormality: psychological effects on women in low-risk pregnancies. Baillieres Best Pract Res Obstet Gynaecol 14:731–747

Statham H, Solomou W, Green J M (2002) When a baby has an abnormality: a study of health professionals' experiences (vol 2 of the report of a study funded by a grant from the NHS (R&D) Mother and Child Health Initiative (MCH 4–12) 'Detection of fetal abnormality at different gestations: impact on parents and service implications'. Cambridge: Centre for Family Research). University of Cambridge

Stern D (1995) The motherhood constellation: a unified view of parent-infant psychotherapy. Basic Books, New York

Stroebe M (1992) Coping with bereavement: a review of the grief work hypothesis. Omega 26:19–42

Toedter LJ, Lasker JN, Janssen HJ (2002) International comparison of studies using the perinatal grief scale: a decade of research on pregnancy loss. Death Stud 25(3):205–228

Wertz DD, Fletcher JC, Nippert G (2002) In focus. Has patient autonomy gone to far? Geneticist's view in 356 nations. Am J Bioeth 2:W21

White-van Mourik M, Connor JM, Ferguson-Smith MA (1992) The psychosocial sequelae of a second-trimester termination of pregnancy for fetal abnormality. Prenat Diagn 12:189–204

Zeanah CH (1989) Adaptation following perinatal loss: a critical review. J Am Acad Child Adolesc Psychiatry 28:467–480

Zeenah CH, Dailey JV, Rosenblatt MJ, Saller DN (1993) Do women grieve after terminating pregnancies because of fetal anomalies? A controlled investigation. Obstet Genecology 82:270–275

Chapter 2
Managing Complex Psychoanalytic Research Projects Applying Mapping Techniques – Using the Example of the EDIG Study

Nicole Pfenning-Meerkötter

Abstract Science and research today are embedded in the development of a knowledge society, implying that scientifically based knowledge gains more and more importance in our society. As the demand for scientifically based knowledge and progress is continually growing, and science poses new risks for society and raises new questions, the need for regulating and controlling scientific activities has grown stronger. This development towards a stronger political and societal impact on research implies that scientists nowadays not only have to meet specific quality criteria, formulated by the scientific communities themselves, but also external criteria formulated by politicians, governments and other third-party funders. The EDIG study, funded by the European Commission, exemplifies this development towards increasingly complex research in a globalized context. In the EDIG project multiple research questions were investigated, applying a broad spectrum of research instruments and methodologies, while engaging in an interdisciplinary dialogue, in order to meet external and internal quality criteria. Graphical tools such as mind maps, flowcharts or concept maps may provide a useful instrument for researchers to meet the complexities and endeavours of current research projects. Such mapping techniques not only support the user in identifying and retrieving knowledge but they may also be used as an externalized memory. In constructing a map, the researcher is forced to reflect systematically on his/her knowledge, to elaborate on it and thereby identify possible knowledge gaps. By making the user's cognitions explicit, knowledge is more easily accessible to the researcher as well as his/her team members, thus enhancing communication processes as well as intersubjective transparency.

Keywords Knowledge management • Knowledge society • Mind map • Psychoanalytic research

---

N. Pfenning-Meerkötter (✉)
Sigmund-Freud-Institut, Myliusstraße 20, Frankfurt 60323, Germany
e-mail: pfenning@sigmund-freud-institut.de

2.1 Challenges in Psychoanalytic Research

An EU research project like the EDIG study poses special challenges for the researchers, calling not only for expert knowledge, as well as methodological competences, but also for profound management skills. The EDIG study, with its broad methodological approach and taking into account interdisciplinary aspects and intercultural differences (cf. Chap. 3), illustrates the challenges that scientific projects have to face nowadays. As the Bielefeld research group around the sociologist Peter Weingart (Weingart et al. 2007) describe, science and research today are embedded in the development of a knowledge society, implying that scientifically based knowledge gains more and more importance in our society. Experts and expert knowledge are in great demand when it comes to the formulation of scientifically based recommendations for legal, societal and political frameworks for future policies. Whereas scientific knowledge increasingly replaces 'naïve' notions, and thereby contributes to the development of new insights in all domains of research activity, it also continuously creates new problems and raises new questions. As research is forced to explain increasingly complex phenomena in very different contexts, this leads to heightened insecurity in the predictive power and possible consequences of scientific results and developments (as shown in this volume; Weingart et al. 2007).

As the demand for scientifically based knowledge and progress is continually growing, and science poses new risks for society and raises new questions, the need for regulating and controlling scientific activities has grown stronger. This development towards a stronger political and societal impact on research implies that scientists nowadays not only have to meet specific quality criteria, formulated by the scientific communities themselves, but also external criteria formulated by politicians, governments and other third-party funders. Therefore, when evaluating research grants, not only the axiomatic relevance of a research endeavour is taken into account, but also its technological and societal impact (Carrier et al. 2007).

The Sigmund-Freud-Institute, a traditional institute for psychoanalytic research in Germany, is also affected by this development. The EDIG study, funded by the European Commission, exemplifies this development towards increasingly complex research in a globalized context. It took up a concern of the European Commission, formulated in one of their research work programmes "Science and Society", located within the 6th EU Framework Programme for Research and Technological Development (FP6). One of their main focuses was to support a responsible use of scientific and technological progress, including research on ethics in relation to science. Hence, a crucial criterion for evaluating research grants was the extent to which "[…] the proposed project is likely to have an impact on reinforcing competitiveness or on solving societal problems." (European Commission 2003, p. 32).

However, the European Commission not only considered the practical output of a prospective research project but also its innovative power. A highly innovative aspect of the EDIG study was to connect the expert knowledge of different disciplines, such as psychoanalysis, ethics, medicine as well as social sciences. Prior to 2003, there had been no detailed analysis of the relation between ethical reflections in the decision-making processes around prenatal diagnosis (PND) and psychic strains on the individuals involved (Leuzinger-Bohleber et al. 2008).

By the conception of such a study the consortium of researchers also reacted to a desideratum of the European Commission, which sought to promote the emergence of structural connections, dialogues and networks within the European Research Area (European Commission 2003). This approach, of supporting cooperation between different disciplines, resulted from the perception that excellent research is characterized more and more by its complexity and its interdisciplinary nature (European Commission 2002).

Another specific characteristic of the EDIG study was its multinational approach with the participation of Italy, Sweden, Israel, Greece, England and Germany. According to the European Commission, the necessity to cooperate on a cross-national level, within Europe but also world wide, derives from the demands that arise in a globalized environment. The aim of the Sixth Framework Programme was thus to encourage the emergence of a European Research Area, promoting the establishment of common European policies (European Commission 2003). This demand was met by the efforts of the EDIG study to include a variety of countries, characterized by their diversity regarding law, policy and practice of prenatal testing.

Of course, the EDIG study also had to prove itself on a scientific-axiomatic level, showing its methodological soundness. Psychoanalytic research has to deal with the ideal of a 'unified' conceptualization of science and scientific methods and is critically observed by non-psychoanalytical scientists as well as psychoanalysts. Considering this background, the development of a study design, which adequately addresses unconscious processes and fantasies, proved to be a particular challenge. To meet these new challenges, Leuzinger-Bohleber, the coordinator of the EDIG study, in line with a view held by many German psychoanalysts, developed a specific approach to psychoanalytic research. This view is marked by its striving for intersubjectivity, trying to integrate and reflect different types of investigations, methodologies and instruments (cf. e.g. Leuzinger-Bohleber 2002, 2007; Leuzinger-Bohleber and Fischmann 2006).

2.2 Impact on the Research Process

As shall be shown in the following sections, these external and internal quality criteria and the specific approach to psychoanalytic research as outlined had a marked impact on the research process, increasing its complexity. The project's complexity resulted from an integration of several project objectives and research questions, the combination of different forms of research (clinical and extra-clinical; empirical and interdisciplinary) and the interplay of various methodological approaches, including the application of psychoanalytic and non-psychoanalytic instruments, quantitative and qualitative data, bottom-up and top-down-processes as well as nomothetic, group-statistical and idiosyncratic analyses on a single case level. The specific research approach thus affected the research process, including the theoretical preparatory work, the performance of the study, data analysis and dissemination (Bortz and Döring 2006).

By referring to linear models depicting the research process in the life sciences, e.g. one developed by Bortz and Döring (2006), the problems arising due to the complexity of the research approach, will be shown. Such linear models are always restricted to extracts of an iterative scientific process and do not meet up with the dynamic character of research actions. However, choosing a schematic, phase-oriented approach offers the advantage of depicting the complex research process in all its facets.

Usually, a research project starts with the preparatory phase, followed by the actual study performance, then the data analysis and finally the dissemination of study results. The preparatory phase for a research project itself incorporates different steps. In many cases it starts with a comprehensive literature review, the specification of relevant research questions and the development of a sound study design – including the specification of the type of investigation, the operationalization of the relevant concepts and variables as well as the population to be investigated. This is followed by the conceptualization of the study performance as well as the anticipation of the data analysis (Bortz and Döring 2006). These different phases will now be described in detail.

2.2.1 Literature Review

Firstly, a research process should include a comprehensive literature review. In the sixties, de Solla Price, a historian of science and physicist, investigated the growth of natural sciences and concluded that the growth in the last two centuries resembles an exponential curve with a duplication period of only 10–15 years (de Solla Price 1974). Although criticized for his methodological approach, there is broad consensus that scientific activity, including the number of publications as an output-criterion, has decidedly grown in the last decades (cf. e. g. Kölbel 2004; Weingart 2005).

Looking at a research project like the EDIG study, with multiple objectives, intercultural comparisons and multidisciplinary research questions – and in a field of research that attracts high scientific interest – the complexity of a comprehensive literature review quickly becomes apparent. This is exacerbated by the fact that the scientific activity, including the number of publications, has grown enormously in the last decades. Processing this amount of knowledge can easily overload individuals' information processing capacities. Moreover, the interdisciplinary approach made it necessary to connect the expert knowledge of different disciplines.

2.2.2 Identification of Relevant Research Questions

Closely related to the question of literature research is the identification of relevant research questions. They might result from a research assignment, from practical/societal problems that elicit the need for scientific solutions or they might be based on a "purely" scientific interest (Friedrichs 1990). The EDIG project was confronted with the high standards of the European Commission and sought to address not only theoretical questions but also to contribute to the solution of societal problems. Thus it had to keep track of multiple research questions that split into further aspects. Moreover, all findings had to be highlighted from an intercultural and interdisciplinary perspective, increasing the number of relevant research questions further.

When looking at the research process, one realizes that it often shows dynamic characteristics and can be described in terms of oscillating processes, which imply the modification of theories, methods and data models. Overlaps, interferences, leaps, feedback and loops are regular processes (Hug 2001; Kromrey 2006). Therefore, the project planning has to be flexible enough to allow for changes and revision. Furthermore, interesting research questions sometimes do not emerge until the study has started (Diekmann 2006). Therefore, in the EDIG study, a heuristic part, consisting of interviews with pregnant women and their partners, was included to promote the development of preliminary hypotheses on the processing of ethical dilemmas in the context of prenatal testing. Considering the fact that each of these interrelated decisions is crucial with regard to the research design, Kromrey (2006) recommends a careful documentation of changes, amendments etc. in the research process to ensure intersubjective transparency.

2.2.3 Defining the Type of Investigation

After reviewing the current theoretical and empirical status, and having specified the relevant research questions, the researcher is asked to define the type of investigation which offers an adequate methodological approach to the research question. Bortz and Döring (2006) differentiate between descriptive studies, explorative studies and studies which test hypotheses. The EDIG study combined all aspects: first, the study aimed to describe the populations in the different participating countries that use the techniques of prenatal testing. Furthermore, it was characterized by its combination of bottom-up or inductive with top-down or deductive processes. On the one hand, hypotheses were formulated, based on psychoanalytic and psychological knowledge as well as existing empirical studies in the field, such as the investigation by Statham et al. (2002, 2003). On the other hand, the interviews conducted with couples undertaking PND, as well as psychoanalysts and their former patients, served as a heuristic basis for formulating the first tentative hypotheses. Moreover, group differences, correlations, changes over time, as well as single case studies, were incorporated within the study design. Thus, the combination of these different types of investigations caused a complex interplay of descriptive, inductive and deductive research processes.

2.2.4 Definition of Concepts, Variables and Operationalization

After defining relevant research questions, as well as the type of investigation and study design, the central variables have to be set and operationalized. With the constant growth of our knowledge base, science continuously creates new scientific questions, forcing research to solve increasingly complex problems (Carrier et al. 2007). Authors like Rescher (1982) argue that efforts (methods, manpower, technologies) have to be continuously increased in order to achieve the same relevant results. The EDIG project tried to contribute to the standard of knowledge by choosing an interdisciplinary approach, bringing together disciplines that had not previously been involved in a multidisciplinary dialogue in the field of PND – namely psychoanalysis, cooperating with ethics and medicine. Moreover, up to that point longitudinal investigations of couples undergoing PND, from the time of waiting for the test results up to the expected date of birth and beyond, were clearly missing from previous literature.

These considerations resulted in the inclusion of a large number of variables within the study in order to meet the relevant research objectives and questions, e.g. to investigate the relation between a number of predictors and different dependent variables. Moreover, careful operationalization was required that would allow for the assessment of conscious as well as unconscious processes in a methodologically sound way. In the EDIG project standardized scales were used, as well as multiple choice questions with given answers, open questions and interviews, resulting in a large pool of quantitative and qualitative data.

2.2.5 Specification of Subjects

In the EDIG study, the population consisted of couples choosing prenatal testing (substudy A) as well as psychoanalysts, whose former patients reported retrospectively on their experience of PND (substudy B).

2.2.6 Conceptualization of Study Performance and Data Analysis

In this preparatory phase the performance of the study has to be anticipated as well as the analysis of data. This implies decisions about the processing of raw data (e.g. of open questions) as well as choosing an adequate statistical procedure for data analysis. Of course, the high number of research questions and associated variables entail complex statistical analyses in order to investigate correlations, group differences or changes over time. Moreover, the question arises of how to relate the different data sources to each other, e.g. how material gained from open questions could be related to the standardized scales – a question which touches the aspect of methodological triangulation.

2.2.7 The Performance of the Study

After having defined the study design, the study has to begin. As can be imagined, the performance of such a research project requires high organizational efforts. Hence, special skills in the field of project management are required, as Bulmahn, the former federal minister for education and research in Germany put it (Bundesministerium für Bildung und Forschung 2002). Collaboration with colleagues, which is indispensable for the realization of studies of such scope, presupposes efficient communication.

2.2.8 Data Analysis

As soon as the data has been collected, the data analysis has to begin. It is a well-known fact that research interests often stimulate the formulation of manifold hypotheses that, due to limited capacities, cannot be processed simultaneously. In such cases, the research reality makes it necessary to determine strategies for data analysis, i.e. to fix priorities and proceed successively. In some cases, the "data flood" might lead to an overload, ending up in "data graveyard".

2.2.9 Final Report/Dissemination

In line with the demand of the European Commission to heighten the acceptance of science and technology by fostering a dialogue between scientists and the public (cf. work programme "Science and Society"; European Commission 2003), the dissemination of the results was a crucial factor. By initiating a public debate on the topic of ethical dilemmas due to prenatal testing, the EDIG study consortium tried to meet one of the evaluation criteria of the work programme, namely the readiness to engage with actors beyond the research community and with the public as a whole (Council Regulations on the Rules for Participation, article 10; Europäisches Parlament und Rat der Europäischen Union 2002). However, the interdisciplinary and intercultural approach of the study made it difficult to spread the results within very different scientific communities and diverse institutions, engaged in the field of prenatal testing, such as medical doctors, psychologists, psychoanalysts, ethicists and politicians.

2.2.10 Summary

Accordingly, the delineated research approach had particular consequences for the overall research process, influencing the conceptualization and preparatory work, the performance of the study, the data analysis and the dissemination of the results. The EDIG study had to face challenging tasks. The growing number of relevant publications had to be identified in order to determine relevant research questions. The complex study design (longitudinal study, large study population, methodological triangulation, high number of relevant concepts and variables, different research objectives and methodological approaches) entailed methodological and organisational challenges. Moreover, the knowledge of the various experts engaged in the interdisciplinary and cross-cultural dialogue had to be integrated.

Dealing with such complexities can be challenging. The production of knowledge, as realized in a research project like the EDIG study, calls for supporting tools to help manage these requirements. Nowadays, the quality and success of problem solving processes proves more than ever to be dependent upon the capacity to structure complex knowledge, to communicate and to keep it available for use.

Visualization strategies are seen as a key to the successful handling of information and knowledge. In the last 30 years, within psychology and educational sciences, a group of visualization tools have been developed which proved to be very supportive for knowledge management, namely mapping techniques (Jonassen et al. 1993; Mandl and Fischer 2000; Tergan 2004).

2.3 Mapping Techniques

Mapping tools allow for a structured, spatial-visual representation of knowledge units. Mind maps, originally developed by Tony Buzan (Buzan and Buzan 1999), represent one form of mapping technique. When constructing a mind map, the central theme, represented by a word, picture etc. is placed in the middle. The main topics of the subject radiate from the centre like branches, which are labelled using primary ideas. Topics of subordinate meaning are portrayed as branches, connected to higher order limbs. These are labelled with secondary or tertiary ideas, building a structure of connected nodal points (Haller 2002). The underlying assumption is that the accomplishment of cognitively challenging situations can be alleviated by making those cognitions that form the knowledge basis explicit through visualization techniques (Tergan 2004; Hillen et al. 2000; Hardy and Stadelhofer 2006; Hillen et al. 2000; Tergan 2004). A similar yet more complex mapping technique was developed by Joseph D. Novak (Novak and Gowin 1984), called concept maps. Concept maps include concepts, usually represented within circles or boxes, and relationships between concepts, indicated by connecting lines between concepts. In contrast to mind maps, these lines are explicitly labelled, specifying the relationship between the two concepts.

Mapping techniques are useful tools to promote reflection by the user. They help to structure complex informational settings and to relate information units, where necessary (Thüring et al. 1995). Choosing a spatial-visual approach enables the user to take advantage of his/her abilities of spatial orientation and creation of cognitive maps in order to orient in "knowledge space". The analogue modelling of structure helps to capture the overall structure immediately and thereby acquire a first framework, which serves to orient oneself in the presented contents. This kind of information editing is highly relevant for comprehensive topics, as the processing of complex data requires that many information units and their mutual connections are kept active in working memory. Since our memory capacities are limited, an external representation of information units discharges memory. The newly acquired resources can be used for elaboration and development of further considerations (Haller 2002).

Computer-based mapping techniques have proven to be of particular use when managing knowledge. They allow for a link-up with external knowledge databases such as thesauri, encyclopaedias, databases, internet files or private files. Such link-ups facilitate the individual organization of knowledge and information by setting up personal deposits (Haller 2002; Reinmann-Rothmeier and Mandl 2000; Tergan 2004).

2.3.1 Mapping in the EDIG Study

In the EDIG study different tools, supporting the process of knowledge management, were applied in the different parts of the research process. In the next section the research process, with its associated difficulties, and the instruments used to address these difficulties shall be denoted.

2.3.1.1 Literature Research

As mentioned above, the topic of prenatal testing and associated ethical dilemmas is well researched, resulting in a large number of publications, which easily overload one's memory capacities. Moreover, the existing expert knowledge had to be connected (cf. Sect. 2.2.1).

To avoid memory overload and to ensure an exchange of expert knowledge, the knowledge had to be externalized and represented in some form. For that, we built up an electronic, online-literature database, comprising bibliographic information on publications relevant to the EDIG study. Moreover, in Frankfurt, a knowledge management software was applied, called Citavi (www.citavi.com), supporting the user to organize and structure knowledge on one specific topic.

2.3.1.2 Identification of Relevant Research Questions

As described before, the researchers had to keep track of numerous research questions. In order to manage these we took advantage of the benefits of mapping techniques. In doing so, we followed a specific procedure. First, the various research questions were identified and listed, derived from the EU-application for the study and theoretical considerations and taking into account existing empirical studies, such as the "Cambridge-Study" by Statham, Solomou and Green (Statham et al. 2002; Statham et al. 2003) as well as other current publications on the subject of prenatal diagnosis. Moreover, experts in this field were consulted on recent empirical findings and publication activities in this research field. Major topics comprised:

- the description of the sample;
- the way prenatal testing was experienced and processed;
- the investigation of ethical dilemmas in the context of prenatal testing, taking up the interdisciplinary approach of the EDIG study;
- and finally, the aspect of counselling/crisis intervention, reflecting the aim of the EDIG study to contribute to quality standards for counselling in the field of PND.

Each major branch comprised more specific research questions: e.g., when looking at the way, PND was experienced and processed, the questions arose:

- What reasons were given for having PND?
- Which attitude towards PND proved to be characteristic for this population?

All these aspects were then visualized by using the mind mapping technique (cf. Fig. 2.1).

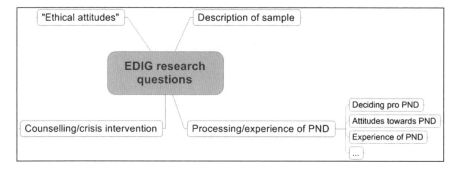

Fig. 2.1 Mind map (excerpt) of relevant research questions in the EDIG project

By explicitly formulating the research questions using a graphical tool for knowledge representation, an active process of reflection was initiated, enabling the user to identify and record the major aims of the study. Given the multitude of research questions that were relevant to the project, the visualizing technique was chosen as an adequate way to handle this complex information. Again, the formulation of research questions requires expertise in a scientific field in order to identify research demands. By representing information in the form of a mind map, the user's cognitive processes were made explicit and therefore accessible to other experts in the field. By sharing the user's knowledge structure with other experts, possible mistakes or missing research questions could be detected more easily. An example of this is that after constructing the first map, I sent it to Helen Statham, who is an experienced researcher in the field of prenatal testing and asked her to check that no relevant points were forgotten.

2.3.1.3 Defining Type of Investigation

As mentioned before, the EDIG study combined different types of investigation. Managing the complex interplay between bottom-up and top-down processes proved to be a particular challenge. For us, it was very interesting to relate newly derived hypotheses from the heuristic part of the study to other sources of data, e.g. to questionnaires. Therefore, the hypotheses, derived from the psychoanalytic interviews were rephrased as research questions and included in the map, marked as post-hoc-analyses. In this context, the mapping technique ought to support the researcher in organizing and monitoring the interplay of bottom-up and top-down-processes and thereby ensure transparency.

We also had to keep track of correlations, group differences etc. All these aspects were integrated into the map, formulating research questions that address, for example, group differences or time courses. The map suggested how to proceed with data analysis, for example, to first explore the data set, then look at group differences and then investigate correlations. Again, the mind map served the purpose to support the researcher and to structure a systematic approach to analysing the data (cf. Fig. 2.2).

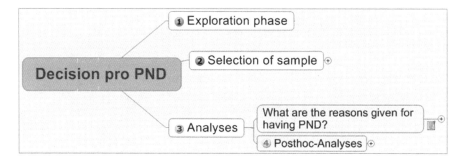

Fig. 2.2 Structure of mind map (excerpt) in the EDIG project

When talking about different types of investigation, one must not forget that the EDIG study not only looked at group statistics but also single cases. This brings up the question of how to combine these different approaches to ethical dilemmas – leading to the next point, the operationalization of concepts and data analysis.

2.3.1.4 Definition of Concepts, Variables and Operationalization

After having listed the research questions, the researcher has to be aware of the relevant variables and the specification of how these will be measured, i.e. operationalizing the variables and concepts. As described before, the research questions were assessed by a broad spectrum of variables and operational definitions, e.g. the attitude towards PND was investigated by using a self-constructed rating-scale, as well as open questions and interviews. This aspect alludes to the topic of data triangulation, which deals with the question of how to combine different sources of data, gained from different methodological approaches, namely quantitative and qualitative approaches. Authors with a constructionist background depict the utility of triangulation in terms of adding a sense of richness and complexity to an inquiry (Flick 1995; Kelle 2001).

The second step of the mapping procedure therefore included identifying the relevant sources of information and fixing them in the map. The possible operational definitions of each research question were listed using so-called annotations – notes, displayed in a separate window. The synopsis of different operationalizations should enable the researcher to highlight research questions from different perspectives for a more holistic and rich approach. Moreover, it should facilitate the reflection which source of information led to what kind of statement. And finally a more structured and organized approach to data analysis should be supported, which supports the researcher (cf. Fig. 2.3).

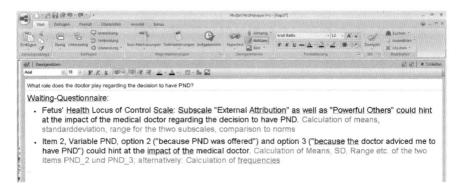

Fig. 2.3 Mind map annotation, listing the relevant operationalizations and statistical analyses with regard to the research question, "What role does the doctor play regarding the decision to have PND?"

2.3.1.5 Specification of Subjects

Having to deal with different subsamples increases the organizational requirements, which affects the performance of the study. Moreover, investigating group differences also increases the complexity of data analysis. The mind map with its organizational structure helped to monitor the different subpopulations and to account for their different analyses by calculation of group differences.

2.3.1.6 Conceptualization of Study Performance and Data Analysis

In a third and last step, suggestions for statistical procedures are included in the annotations, in order to prepare appropriate statistical analyses (e.g. t-tests, analyses of variance, etc.). This is the attempt to make the expert's operational knowledge explicit. Knowledge management should not only refer to explicit analytic knowledge on concepts but also to skills ("Können"), including heuristic strategies (Gruber et al. 2004). The mapping procedure, in this context, allows for the discussion of the user's operational knowledge, making it shareable and adding a group corrective. Moreover, it allows other, less experienced members of the research team to follow and reproduce the assigned steps. Finally, mapping programs like Mindjet allow for the organization of external knowledge resources, e.g. by setting links to Internet documents on complex statistical procedures or files stored on the personal computer, comprising files such as SPSS syntaxes.

2.3.1.7 The Performance of the Study

Of course, much effort is needed to organize, coordinate and perform such a study. Graphical tools may support the conduction of the study, e.g. flowcharts, which can be applied to coordinate the researcher's activity. A flowchart usually represents an algorithm or process, where the successive steps are depicted as boxes and their order by connecting arrows. They are usually applied to analyze, design, document or manage a process or program. By using such a graphical tool, the study's procedure can be made explicit and transparent to the research team, facilitating coordination of the activities of team members. This chart was applied in another study context and proved to be helpful in organizing the researchers' activities.

2.3.1.8 Data Analysis

Due to limited personnel and time it is not always possible to explore each research question simultaneously. The mind map may help the researcher to set priorities and coordinate the analysis of data. Even if limited capacity does not allow the analysis of all aspects at the same time, analyses may be continued at any point when all relevant research questions, associated operationalizations and possible statistical analyses are listed. Moreover, the evolution of new research questions and the performance of post-hoc statistical analyses are carefully documented for intersubjective transparency.

2.3.1.9 Final Report/Dissemination

The dissemination of the EDIG study results proved to be particularly challenging, because very different disciplines, organizations, sections, institutions and persons are engaged in the field of prenatal testing – including medical doctors, ethicists, psychoanalysts, politicians, and support groups among others. Again, a graphical tool was applied to visualize the network between the existing institutions.

2.4 Summary

Graphical tools such as mind maps, flowcharts or concept maps may provide a useful instrument for researchers to meet the complexities and endeavours of current research projects. Such mapping techniques not only support the user in identifying and retrieving knowledge, they may also be used as an externalized memory. In constructing a map, the researcher is forced to reflect systematically on his/her knowledge, to elaborate on it and thereby identify possible knowledge gaps. By making the user's cognitions explicit, knowledge is more easily accessible to the researcher as well as his/her team members, thus enhancing communication processes as well as intersubjective transparency. Of course, such techniques only serve their purpose if the researcher defining the map is trained properly and the whole research team is committed to the underlying concept. However, – as I have tried to delineate in this article – if they are introduced carefully with regard to the needs and demands of the respective research objectives, they may constitute a major benefit for the whole research endeavour.

References

Bortz J, Döring N (2006) Forschungsmethoden und Evaluation für Human- und Sozialwissenschaftler. Springer, Berlin/Heidelberg/New York

Bundesministerium für Bildung und Forschung (2002) Chance für Deutschland und Europa: Das 6. Forschungsrahmenprogramm. http://www.bmbf.de/pub/das_sechste_forschungsrahmenprogramm.pdf.Retrieved 22 May 2008

Buzan T, Buzan B (1999) Das Mind-map-Buch: Die beste Methode zur Steigerung ihres geistigen Potentials, 4th edn. MVG, Landsberg am Lech

Carrier M, Krohn W, Weingart P (2007) Historische Entwicklungen der Wissensordnung und ihre gegenwärtigen Probleme. In: Weingart P, Carrier M, Krohn W (eds.) Nachrichten aus der Wissensgesellschaft. Analysen zur Veränderung der Wissenschaft, 1st edn. Velbrück-Wissenschaft, Weilerswist, pp 11–33

Diekmann A (2006) Empirische Sozialforschung: Grundlagen, Methoden, Anwendungen, 16th edn. Rowohlt-Taschenbuch, Reinbek bei Hamburg (Orig.-Ausg.)

Europäisches Parlament und Rat der Europäischen Union (2002) Verordnug des Europäischen Parlaments und des Rates über Regeln für die Beteiligung von Unternehmen, Forschungszentren und Hochschulen an der Durchführung des Sechsten Rahmenprogramms der Europäischen Gemeinschaft (2002–2006) sowie für die Verbreitung der Forschungsergebnisse. http://eur-lex. europa.eu/LexUriServ/LexUriServ.do?uri=OJ:L:2002:355:0023:0034:de:pdf

European Commission (2002) The sixth framework programme in brief. ec.europa.eu/research/fp6/pdf/fp6-in-brief-en.pdf

European Commission (2003) Work programme: 2003. Structuring the ERA. Science and society (section 4). ftp://ftp.cordis.europa.eu/pub/fp6/docs/wp/sp2/t_wp_200203_en.pdf

Flick U (1995) Triangulation. In: Flick U (ed.) Handbuch qualitative Sozialforschung. Grundlagen, Konzepte, Methoden und Anwendungen, 2nd edn. Beltz, Psychologie Verlag Union, Weinheim, pp 432–434

Friedrichs J (1990) Methoden empirischer Sozialforschung, 14th edn. Westdt, Opladen

Gruber H, Harteis C, Rehrl M (2004) Wissensmanagement und Expertise. In: Reinmann G, Mandl H (eds.) Psychologie des Wissensmanagements. Perspektiven, Theorien und Methoden. Hogrefe, Göttingen, pp 79–88

Haller H (2002) Mappingverfahren zur Wissensorganisation: Diplomarbeit, Freie Universität Berlin, Berlin. http://heikohaller.de/literatur/diplomarbeit/mapping_wissorg_haller.pdf

Hardy I, Stadelhofer B (2006) Concept Maps wirkungsvoll als Strukturierungshilfen einsetzen. Z Pädagog Psychol 20(3):175–187

Hillen S, Berendes K, Breuer K (2000) Systemdynamische Modellbildung als Werkzeug zur Visualisierung, Modellierung und Diagnose von Wissensstrukturen. In: Mandl H, Fischer F (eds.) Wissen sichtbar machen. Wissensmanagement mit Mapping-Techniken. Hogrefe, Göttingen, pp 71–89

Hug T (2001) Einführung in die Forschungsmethodik und Forschungspraxis. Wie kommt Wissenschaft zu Wissen? / Hrsg. von Theo Hug, vol 2. Schneider-Verlag Hohengehren, Baltmannsweiler

Jonassen DH, Beissner K, Yacci M (eds.) (1993) Structural knowledge: techniques for representing, conveying, and acquiring structural knowledge. Lawrence Erlbaum, Hillsdale

Kelle U (2001) Sociological explanations between micro and macro and the integration of qualitative and quantitative methods. Forum Qualitative Sozialforschung/Forum: qualitative social research 2(1):Art. 5. http://nbn-resolving.de/urn:nbn:de:0114-fqs010159

Kölbel M (2004) Wissensmanagement in der Wissenschaft: Das deutsche Wissenschaftssystem und sein Beitrag zur Bewältigung gesellschaftlicher Herausforderungen. Wissenschaftlicher, Berlin

Kromrey H (2006) Empirische Sozialforschung: Modelle und Methoden der standardisierten Datenerhebung und Datenauswertung, 11th edn. Lucius & Lucius, Stuttgart

Leuzinger-Bohleber M (2002) Wissenschaftstheoretische Betrachtungen und methodische Anlage der Studie. In: Leuzinger-Bohleber M, Rüger B, Stuhr U (eds.) "Forschen und heilen" in der Psychoanalyse: Ergebnisse und Berichte aus Forschung und Praxis. Kohlhammer, Stuttgart, pp 10–45

Leuzinger-Bohleber M (2007) Forschende Grundhaltung als abgewerter 'common ground' von psychoanalytischen Praktikern und Forschern? Psyche 61:966–994

Leuzinger-Bohleber M, Fischmann T (2006) What is conceptual research in psychoanalysis? Int J Psychoanal 87:1355–1386

Leuzinger-Bohleber M, Engels E-M, Tsiantis J (2008) Introduction and overview. In: Leuzinger-Bohleber M, Engels E-M, Tsiantis J (eds.) The janus face of prenatal diagnostics. A European study bridging ethics, psychoanalysis and medicine. Karnac, London

Mandl H, Fischer F (eds.) (2000) Wissen sichtbar machen: Wissensmanagement mit Mapping-Techniken. Hogrefe Verlag für Psychologie, Göttingen

Novak JD, Gowin DB (1984) Learning how to learn. University Press, Cambridge

Price, D J de Solla (1974) Little science, big science: Von der Studierstube zur Großforschung, 1st edn. Suhrkamp, Frankfurt am Main

Reinmann-Rothmeier G, Mandl H (2000) Individuelles Wissensmanagement: Strategien für den persönlichen Umgang mit Information und Wissen am Arbeitsplatz, 1st edn. Huber, Bern

Rescher N (1982) Wissenschaftlicher Fortschritt: Eine Studie über die Ökonomie der Forschung. de Gruyter, Berlin

Statham H, Solomou W, Green JM (2002) Emotional well being after a termination for abnormality: the impact of obstetric and social factors. Eur J Hum Genet 10:323–324

Statham H, Solomou W, Green JM (2003) Communication of prenatal screening and diagnosis results to primary-care health professionals. Public Health 117(5):348–357

Tergan S-O (2004) Wissensmanagement mit Concept Maps. In: Reinmann G, Mandl H (eds.) Psychologie des Wissensmanagements: Perspektiven, Theorien und Methoden. Hogrefe, Göttingen, pp 259–266

Thüring M, Hannemann J, Haake J (1995) Hypermedia and cognition: designing for comprehension. Commun ACM 38(8):57–66

Weingart P (2005) Die Stunde der Wahrheit?: Zum Verhältnis der Wissenschaft zu Politik, Wirtschaft und medien in der Wissensgesellschaft. Velbrück-Wissenschaft, Weilerswist

Weingart P, Carrier M, Krohn W (eds.) (2007) Nachrichten aus der Wissensgesellschaft: Analysen zur Veränderung der Wissenschaft, 1st edn. Velbrück-Wissenschaft, Weilerswist

Chapter 3
Distress and Ethical Dilemmas Due to Prenatal and Genetic Diagnostics – Some Empirical Results

Tamara Fischmann

Abstract The empirical results of the EDIG study – an empirical and theoretical study focusing on ethical dilemmas with an interdisciplinary perspective – are depicted here, focusing on distress resulting from modern biotechnology – namely prenatal diagnostics (PND) – and ethical issues stemming from this. Results are presented from the longitudinal investigation of women currently undergoing PND. A brief description of the study design and population is followed by an account of the impact on women of the testing process. The question as to what degree modern technologies pose a threat to the individual is raised and some possible answers suggested. Finally, the central question of this research endeavour, whether or not prenatal testing is perceived as an ethical dilemma or conflict among individual pregnant women will be discussed.

Keywords Prenatal genetic diagnostics • Psychological distress • Decision making • Empirical findings • Ethics

3.1 Overview

The EDIG study is an empirical and theoretical study focusing on ethical dilemmas with an interdisciplinary perspective. In the empirical part of the study two groups of pregnant/expectant couples were compared, those who received negative findings from PND and those who were given positive results. Preliminary questionnaires were used to identify women and their partners who were interested in taking part in the study. Clinical evidence had shown that, even years after a diagnosis, some women (and men) sustain clinical symptoms e.g. depression. This study was especially concerned with identifying characteristics that are associated with the

---

T. Fischmann (✉)
Sigmund-Freud-Institut, Myliusstraße 20, Frankfurt 60323, Germany
e-mail: Dr.Fischmann@sigmund-freud-institut.de

risk of psychiatric illness as a consequence of the diagnostic process or the result. A further goal was to identify factors that might help to minimize the risk of later illness, for example actions that can be taken at an early time point in PND, such as providing comprehensive information, counselling and crisis intervention (see e.g. Leuzinger-Bohleber 2008).

At the centre of the more theoretical perspective is the question of how ethical dilemmas are perceived. Experts from the involved disciplines (i.e. Psychoanalysis, Ethics, and Medicine) agree that a multi-level perspective has to be taken – taking into account biological, social, cultural, personal, developmental and psychological factors – in order to gain new insights into the way ethical dilemmas are perceived and processed. The EDIG study has incorporated measures of many of these factors as well as measures of psychological and psychodynamic variables.

3.2 Empirical Results

Although the EDIG study comprised two parts, the part that will be the focus of this chapter is referred to as substudy A^1 – a prospective longitudinal investigation of women undergoing PND. Empirical results will be presented and discussed after an overview of the study design and information about the study population. A description of the impact of the testing procedure is followed by an analysis on how prenatal testing relates to distress. This raises the question of the degree to which modern technologies pose a threat to the individual – also searching for predictive factors allowing to allocate possible stressors to prevent psychopathological symptoms in the aftermath of PND. I will conclude with a discussion of the central question of this research endeavour, whether or not prenatal testing is perceived as an ethical dilemma or conflict among individual pregnant women.

Data will also be presented concerning possible predictive factors.

3.2.1 Participants

Study participants were recruited from patients having some form of PND (either amniocentesis, CVS or anomaly scanning) at a private obstetric and perinatology practice (day clinic) or a hospital in the countries involved (i.e. Germany, Greece, Italy, Israel and Sweden). The women were asked to participate in the study when they were having the prenatal test and only those who gave written consent were included.

---

[1] For further results see Leuzinger-Bohleber et al. (2008).

3 Distress and Ethical Dilemmas Due to Prenatal and Genetic Diagnostics

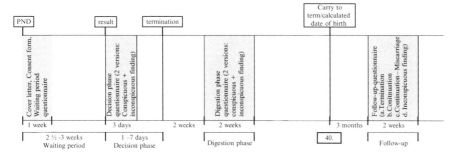

Fig. 3.1 Study design

3.2.2 Study Design

All women were asked to complete a number of questionnaires, comprising standardized instruments as well as non-standardized questions used in other studies and ones devised specifically for this project. This chapter will focus on one outcome measure, the Hospital Anxiety and Depression Scale (HADS), a self-assessment scale developed for detecting states of depression and anxiety in the setting of a hospital medical outpatient clinic. The two subscales, capturing clinically significant anxiety and depressiveness are also valid measures of severity of the emotional disorder, and therefore informative and especially useful when looking for progress through repeated measurement (Zigmond and Snaith 1983). We included the HADS in questionnaires at four consecutive time-points, allowing distress to be measured over the course of the prenatal testing procedure and afterwards. These time points are referred to as 'waiting phase' 'decision phase', 'digestion phase' and 'follow-up'.

Figure 3.1 depicts the phases of the study.

3.2.3 Study Population

One thousand eight hundred and thirty five participants (female and male participants) – were recruited to the study (Germany N = 864, Israel N = 303, Italy N = 368 and Greece N = 300). Seventy percent of women returned questionnaires while waiting for test results (waiting phase) 56.7% in the decision phase 43.9% when assimilating the results (digestion phase) and 29% at the final follow-up phase.

The populations in the different centres were comparable, with a few differences in socio-demographic characteristics, for example Israeli women were younger; Italian women joined the study earlier in pregnancy and Greek women later.

As one main objective of the EDIG study was to explore ethical conflicts and dilemmas in the context of a confirmed genetic and prenatal diagnostics, the study population was subdivided into two groups. Women in group A (N = 82) had received a positive test-result and those in group B (N = 598) a negative result.

3.2.4 Distress – A Uniform Experience for All?

In our study, distress – as measured by HADS-D (Depressiveness) and HADS-A (Anxiety) – takes different courses during the process of PND. For those women with a positive result there is a marked average increase in scores for depression and anxiety after they were told the result. Thereafter there was a marked decline in the level of distress reported by those women with a negative result. It is striking that scores on the HADS remained non-clinical throughout, suggesting that any ethical dilemma, which we hypothesized women might experience in the aftermath of a conspicuous PND-result, does not evoke lasting distress but rather a short and severe episode of anxiety and depression.

However, standard deviations (SD) of scores on the HADS-D and HADS-A ranged between 4.22 and 4.96 for anxiety (mean = 8.43) and 4.58 and 5.18 for depressiveness (M = 6.56). This large variance in the data suggests that distress takes very individual courses and this is illustrated with the following three case-studies.

All three women had the same initial scores for both anxiety and depression scales, but anxiety and depression developed in completely different ways, which, we suggest, was related to the personality structure of the women:

Case Study 1
Mrs. H is 39 years and had tried for a long time to become pregnant. She had two early miscarriages (5th and 12th week) prior to this pregnancy. When she was told that there was a suspicion of a chromosomal disorder (increased nuchal translucency) she was very distressed because she feared that she would lose this baby as well. When the final PND result confirmed the chromosomal disorder to be a Trisomy 21 she was very relieved because she could now choose to have this baby. Her scores declined between the waiting and decision phases. If the diagnosis had been Trisomy 13 or 18 the child would not have been viable. The expert validation (cf. Leuzinger-Bohleber et al. 2008, p. 156 ff) of the interview suggested that Mrs. H is a very determined person, leaving no room for other options than the one of which she is convinced. This coping mechanism seemed to help her cope with her ambivalence. She had decided to be happy because Down syndrome is – as she put it – 'a special and livable condition'.

Case Study 2
Mrs. K is a 36-year-old woman, who was expecting her second child. First-trimester-screening showed low AFP-values and she was advised to have amniocentesis to clarify the suspicion. The final PND result showed that her unborn baby was suffering from Trisomy 18 and she was told that his life expectancy was between a few days or, at the most, a few weeks. She decided to terminate her pregnancy and, as she put it 'my son was stillborn' in the 22nd week of pregnancy and buried a month later'. After the termination, she speaks about the rollercoaster of feelings she has had afterwards: 'This all cannot be true. My head tells me that the decision was the right one for all of us. However, one question always remains, namely what would have happened if we had decided to let our little one decide by

himself on his day of death...'. The amount of distress that these thoughts caused her expresses itself in the high scores on the depression-scale (HADS-D: 12) during the digestion-phase suggesting a depressive mode of processing what seems to be a subject of guilt. After the termination she was engaged in an intense mourning-process, regularly visiting the grave. Her anxiety scores remained clinically high until the digestion-phase – 6 weeks after testing. Because Mrs K did not wish to be interviewed, our hypotheses are drawn only from the information we have from the questionnaires and thus are more tentative.

Case Study 3
Mrs. J. is 28 years old and pregnant with her first child. The First-Trimester-Screening showed a suspicion, which was confirmed following amniocentesis. Turner syndrome (i.e. molecular detection of Karyotype 45XO) was diagnosed, which Mrs K interpreted as 'her baby being a female without ever being able to become a woman'. 'I was told' she says 'that I can either have the baby, give it up for adoption or abort it'. She decides to have a termination, which she had in the 15th week of her pregnancy. She returned to work very quickly – only a week after the termination. In the digestion-phase, she states that 'her decision was the right one for her and their life, but she feels guilty sometimes'. These guilt feelings do not appear to cause a depressive reaction (HADS-D) but anxiety rises almost into a panic (HADS-A decision >14). Her returning to work very quickly after the termination could be an indicator for a manic defense, trying to deny what really happened by turning to everyday life as if nothing has happened. What supports this assumption is her saying when asked if there is anything about her current circumstances that is difficult for her she says: 'Only that, after a follow-up visit at my gynecologist, where he told me that not everything has been evacuated in the curettage, ... I asked myself if everything is really removed or do I have to have another curettage'. We assume that this could indicate a worrying fantasy that part of a dead body is still in her that could turn against her. She gives further support for this assumption during follow-up saying that she has 'recuperated well and everything has healed well again. Apart from that, I have I allowed myself to enjoy life still a little bit. Now I am pregnant again'. She continues in the next question (how do you feel at the moment?) 'Good, except for the fear of a sick or handicapped child'. Later in the same questionnaire she says when asked how she feels about the termination now: 'Now I have doubts about the diagnosis. I sometimes ask myself if they really diagnosed it correctly, if I have killed this child even though it was healthy? I am unsure.' Her anxiety remains on a mediocre level.

As illustrated here, the cases are very individual and show great variation – psychologically as well as in the empirical data. A cross-sectional analysis (means of HADS-D and -A) of the data of these three cases would not have revealed this range of reactions. On the contrary these single case studies show that the generalizing assumption that biotechnology per se imposes distress does not withstand a closer inspection. It is rather the situation of each woman with her particular background and personal characteristics that determines the amount of distress experienced and her way of coping with it.

3.2.5 *Ethical Aspects*

The following part focuses on the ethical part of our study. At the beginning of this research endeavour, the ethicists posed some ethics questions they wished to investigate empirically. Such a study had not previously been undertaken. The empirical researchers, with backgrounds in psychoanalysis and medicine had first to clarify and understand the terms used by the ethicists. Secondly they had to develop constructs which could serve as means to elucidate appropriate responses. So first, some elucidation to the subject in question will be given. This will be followed by some empirical results.

3.2.5.1 Ethical Dilemmas

An ethical dilemma is defined in the *Cambridge Dictionary of Philosophy* (2001) as a situation where an agent has a strong moral obligation or requirement to adopt each of two alternatives, and neither is overridden, but the agent cannot adopt both alternatives.

The knowledge of risks of pediatric disorders and disabilities, and of the possibilities that PND offers, increasingly leads to difficult situations in decision-making (German Society of Human Genetics 1993). It raises questions about which type of knowledge can be acquired through PND and the consequences associated with it. The test result itself can bring an end to overwhelming worries; it may allow, in rare circumstances, the opportunity for prenatal therapy after the detection of a fetal abnormality; or more frequently it can allow the option to plan postnatal care of a baby with abnormalities. However, it can lead to knowledge which is in itself is a burden, that the baby will have serious abnormalities with no possibility of treatment of the diagnosed condition (Honnefelder 2000), thus often leaving termination as the ultimate choice. Baldus (2006) describes in her investigation of women who were confronted with a trisomy 21 test result where even those who were quite decided on their values experienced a serious decisional conflict no matter what their choice was – termination or carrying to term (cf. Leuzinger-Bohleber et al. 2008, chapter 3.2 and chapter 3.5). The following section describes how participants reacted to positive diagnoses, giving a unique insight into what participants were confronted with when facing this decision.

Eighty-two women in subgroup A were told of positive findings 55 women (67%) decided on termination. Some pregnancies were terminated after the diagnosis of lethal abnormalities and others where the baby would be viable.

Which Conditions Should Be Eligible for Termination and When?

In society, there is a discussion about situations in which termination of pregnancy should be allowed. In the follow-up-questionnaire, participants were asked about the situations for which they thought termination should be available. Strongest support for the availability of termination was indicated when a 'woman's life was in danger' (overall 92.1%) followed by if 'pregnancy resulted from rape or incest' (overall 82.8%) and if the 'infant had a chromosomal disorder' (overall 72.6%). There were no significant differences in attitudes between women who had received a positive result from PND and those who had not. All women were less supportive of the availability of abortion after a prenatal diagnosis of mental (64.7% in favour) or physical (43.4%) disabilities. Even among the 25 women who had been given a positive result, only 12 of those who terminated their pregnancy (48%) agreed that abortion should be available in the situation when a child would be born with a physical disability. Women who continued pregnancies were less likely to support the availability of abortion. There were some differences between women in different countries. In Germany three quarters of women (74.9%; N = 104) agreed that abortion should be available up until the 12th week of pregnancy with only 2.2% (N = 3) supportive of the availability of abortion after the 6th month of pregnancy. In contrast Israeli women are the most liberal regarding the availability of late termination, with one quarter of participants supportive (see Fig. 3.2). Greek women, representing a mixed population of Christians and Muslims, had the most conservative attitudes towards late abortion.

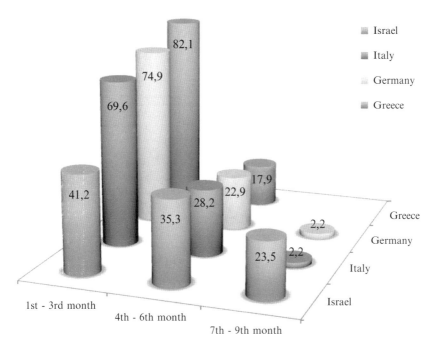

Fig. 3.2 Month of pregnancy until which termination should be allowed (Greece N = 28; Germany N = 139; Italy N = 46; Israel N = 17)

Ethical Issues – Attitudes Towards the Developing Fetus and Acquisition of Human Dignity

In this study we have sought to understand ethical issues from the perspectives of women who had recently given birth or terminated a pregnancy. At the follow-up-phase two questions were posed: what they consider the developing baby (fetus) to be and when they think the developing fetus acquires human dignity. A range of response options were given and women responded to each saying how much she agreed with the given statement.

In general, participants thought a developing baby or fetus was 'a growing baby' with 96% of women with a negative result and 79% of those with a positive result indicating much or very much. Similarly 84% of women with a negative result and 82% of those with a positive result indicated that the developing baby/fetus was 'a human being'. However, this was a difficult question for women to answer, as can be seen from that 38 women (14%) with a negative and another 7 (21%) with a positive result answered this question by stating that they thought the developing fetus to be a "human being" and also said that they thought it to be "biological material". This highlights the dilemma women are in.

In answer to the question about the acquisition of human dignity, only 38% of women (93) with negative results and respectively 40% (12 women) with positive results indicated that a developing baby acquires human dignity 'at conception'. The remaining response options were a gradual evolution from 'development of certain functions', to being 'viable on its own'. Sixteen women with a positive PND-result (62%) and 176 women with a negative PND-result (73%) thought the developing fetus acquires human dignity when it is 'viable on its own'. Fifteen women (56%) with a positive test-result thought it acquires human dignity when it 'developed certain functions', whereas 104 of those with negative PND results (51%) thought this. Finally, 55% of women (17) with positive findings and 71% (166 women) of those with a negative result thought that a fetus acquires human dignity 'with birth'. To summarize: while the majority of both groups of women – those with a negative and those with a positive result – agree that the developing fetus is 'a growing baby' and a 'human being', women with a negative test are more likely to indicate that a developing baby acquires human dignity when it is 'viable on its own' and 'with birth'. Women with a positive result tended to choose when it is 'viable on its own' and when it 'develops certain functions', i.e. the baby acquiring human dignity at an earlier developmental stage.

3.3 Summary

The empirical data reported here derives from the prospective part of the study (Substudy A) studying ethical conflicts and dilemmas in the context of genetic and prenatal diagnostics.

This longitudinal study confirms previous findings (Iles and Gath 1993; Salvesen et al. 1997; Zeanah et al. 1993) concerning distress as a consequence of intentional termination. The assumed ethical dilemma, which women experience in the aftermath of a positive PND result, does not evoke lasting distress but rather a short severe outburst of anxiety and some degree of depressiveness. Distress as measured in our study showed great variance, pointing to the individual courses it takes for different women and questioning the assumption that biotechnology per se imposes distress. Our empirical data suggests it may be personality characteristics of individuals that determine the amount of distress experienced and ways of coping. This is an aspect that should be taken into account when counseling during the process of PND. Similarly, with respect to screening tests, our data suggested (cf. Fischmann in print) that participants were preoccupied by an anticipated PND result, clouding the distress caused by the previous suspicious screening result. Again, the data suggest that a false alarm does not have a distressing effect per se, but depends on the individual and her processing capacities (i.e. defense mechanisms) as a negative test result brings the wished for relief and anxiety diminishes thereafter, although other studies in the literature suggest ongoing residual anxiety (Green et al. 2004).

Within our study we observed 82 women with positive PND results. Of these, 55 women terminated pregnancies (67.1%), some after the detection of viable abnormalities.

As for ethical issues, over three-quarters of our participants thought the developing baby to be a human being that acquired human dignity when it becomes a viable being on its own. Worth mentioning here is that women with positive PND-results had more of a tendency than women with a negative PND result to consider the fetus to be biological material. It is possible that the experience of having received a distressing test-result may affect ethical attitudes. It was not possible to compare cultural differences among women with positive test results because of low numbers overall. A comparison of women with negative results in different countries revealed highly significant differences regarding acquisition of human dignity (cf. Fischmann in print). Here Italian women were more likely to indicate that a developing baby acquires human dignity with conception, whereas the other participants reported the view that human dignity is acquired with birth and when the growing fetus can live on its own. Israeli women were most likely to think that a developing baby acquires human dignity with birth and that a developing fetus is not yet a human being. All this data elicited in our Study reflects some of the cultural differences concerning ethical questions that may be found in these countries.

3.4 Outlook

Investigating ethical conflicts in PND was the first aim of the EDIG study. What are possible ethical conflicts in prenatal diagnostics? A conflict arising from PND may be the dilemma of a pregnant woman when confronted with a problematic test result. This will put her into a position where her wish to become the parent of a mentally, cognitively and bodily healthy child is confronted with a result suggesting something different. The decision to terminate or opt to live with a severely disabled child may be a conflicting experience, raising unconscious murderous fantasies with possible long-term impact, depending on the protective defences available to her. Reasons why parents come to their specific decision is determined by the parents' compassion and perceptions of suffering with their future child, his agony as predicted by his expected impairments, and his or her quality of life. However, quality of life is difficult to define, as it is experienced, lived and judged differently by each and everyone.

From an ethical point of view, biotechnologies evoke both a shift of natural borders and a crossing of natural borders, sometimes transforming biotechnology into a technique of selection, forcing those affected by it to make decisions. As Engels (Leuzinger-Bohleber et al. 2008, p. 255) says, there is "no choice not to choose". Moral conflicts and dilemmas may arise from the "simple" decision to have or not to have PND or for or against an abortion. To give just one example, an individual's autonomous decision to have a disabled child may conflict with a lack of tolerance for people with disabilities in society. This touches our self-concept and moral identity, making it sometimes impossible to come to a decision that will not collide with societal principles. Autonomous decisions by women about their bodies and what they will allow to grow in it collide with different religious and philosophical views on the moral status of the fetus.

Even those who are not religious will be influenced by those views. The question of when a fetus is considered a human being, with all the rights of human beings already born, will necessarily come into play when having to make decisions.

From a Jewish legal point of view, avoiding or mitigating suffering has priority when conflicting ethical interests are considered, making abortion, and even late abortion, an option; the mother's life has priority – she is already born – over that of her unborn child. But, according to the same Jewish law, abortion is considered obsolete if chosen out of fear or concern of what the child might have to endure during life, as life quality is one of the "secrets of God" and not ours to decide upon.

In Christianity the question of ensoulment is at the centre of considerations on the status of the fetus or of abortion. The Vatican has a twofold argument against abortion. One argument is the one of uncertainty of ensoulment from the moment of conception, risking killing an ensouled being through abortion. The other argument is about potentiality, i.e. the potential to become ensouled is there from the moment of conception and should not be violated. This being the predominant Christian view, some moral theologians hold against it arguing that although the death of a fetus at the earliest stage of development remains a loss, sometimes abortion is justified if no "human relationship" can be expected after a severely disabled infant is born. Those theologians demand that the question of what kind of life an infant will have should be at the core of decisions on abortion and not the existence or non-existence of vital life signs. This stands in marked contrast not only to the Vatican but also to Judaism (South Africa).

The Islamic view also accentuates ensoulment, according the fetus with human rights although at a somewhat later time-point than Christianity and Judaism – namely at 120 days after conception as opposed to 40 days after conception.

Abortion in Islam is allowed after implantation and prior to ensoulment in cases of juridical or medical reasons – mainly if the mother's life is endangered.

From a utilitarian point of view the woman is more central to the debate. Refusing a woman an abortion might have more morally intolerable consequences than infanticide or abortion. The comparison between the use of contraceptives and abortion is made in utilitarianism, clearly making a pro-abortion argument ultimately to avoid suffering and frustrated desires. There are only two reasons favoring continuation of a pregnancy with a massively disabled fetus and that is if the termination would sadden the parents or someone would happily adopt such a child.

From a deontological point of view, some hold the position that the fetus lacks any rights to life as long as it lacks consciousness, while the mother already is in possession of this consciousness entitling her to a right to autonomy, which implies a right to abortion.

Depending on which of these views is adopted, decisions on PND or abortion will be made differently and it can be anticipated that coping with the decisions made will also be influenced by it. One can assume psychosomatic reactions especially where specific expectations in PND are frustrated. As PND is supposed to diminish perinatal mortality and morbidity, pregnant women and their partners expect PND to diminish their insecurity of pregnancy and make their dream of a healthy child become a reality – or at least to reduce their anxiety that their future child would not be healthy. Psychoanalytically the expectation that PND will confirm the health of the unborn baby can be interpreted as fending off anxiety and guilt-feelings, which protects the pregnant woman in her current endangered ability to relate during a "tentative pregnancy" for the price of misjudging reality. In case of a negative test result, this scheme will be gratified, but it will result in shock when the test result is a positive finding. Serious decisions will have to be taken, the time of hopeful expectation ends abruptly and confrontation with the subject of life and death is inevitable and guilt-feelings will play a major role. At the time when a decision has to be made, a relationship to the fetus is usually established and as Alexander (1998) showed, couples feel guilty about the death of their baby that they have decided upon or that they had brought about.

Our empirical data on decision-making showed that health professionals are involved in the decision to have PND or not by counselling and not by being the ones making the decision, thus assuring women and their partners an autonomous decision. Hence, the procedure of PND is valued as a relief and accepted by the majority of participants as part of pregnancy, and welcomed for the additional information it provides. Nevertheless, our data also showed that prenatal testing represents an incisive experience, changing the view of pregnancy, where a strong feeling of relief is associated with a negative result and associated with increased uncertainty when the test result is positive, making participants strive for even more information. A positive result evokes distress in the form of a short but severe outburst of anxiety and some degree of depressiveness. The form this distress expresses itself is very individual, depending on the personal situation as much as on personality structures ultimately determining the ability to cope. Once a positive test result is established two thirds of participants in the EDIG study decided to terminate their pregnancy. Crucial for their decisions were such aspects as quality of life of their baby, but also of their own life when envisaging their life with a disabled child. This result informs ethicists on what is important to women when deciding on such sensitive issues as termination of pregnancy. It seems that women are concerned with issues which are much more in touch with reality, like "quality of life". They do not seem to consider societal reactions to their decision; neither are they influenced by religious beliefs, nor the expected amount of social support. On the other hand, the anticipated quality of life of their unborn baby is not an entirely valid argument to approve abortion, the "ranking" within this category goes from chromosomal disorder – where about 70% agree that this is a good reason to justify abortion, – to mental disability (about 65%) to physical disability (about 40%). The vast majority of women endorse abortion if the woman's life was in danger (90%), or if a pregnancy resulted from rape (80%) – unknowingly – supporting a Jewish legal point of view. Women's right to autonomy surprisingly is not an argument of equal value to support abortion. Only about 40% agreed that termination should be available to women who do not want to have a child and only 17% said that abortion should be available to a woman who cannot afford to raise a child. This insight on decision-making in PND shows the influence of personal experience on ethical issues. Although cultural influences were identifiable in the arguments women made their decisions were apparently made according to their personal situation.

This information is important for care giving and counselling matters in PND. In this respect another result of our study is of interest as it revealed that women who were confronted with a positive test result experienced this ambivalently and sometimes even traumatically. Women who decided to terminate their pregnancy after having been informed of a positive result had significantly more intrusive thoughts, were hyper aroused and more showed avoiding behavior afterwards than those who decided to continue their pregnancy. Furthermore our data showed that women who experience feelings of guilt, consisting of self-blame for not having done enough to avoid the loss of the fetus and a sense of personal failure will more likely have depressive reactions, and will be more anxious regarding future pregnancies. In other words, this experience could be a starting point for a pathological mourning reaction. Considering this, it is important to give adequate counseling and care to those involved as a preventive measure of a pathological mourning reaction.

The identification of predictors of distress was another major aim of the EDIG study. Finding such predictors after a problematic PND result may help to optimize the process of PND. The provision of optimal surroundings and care giving may help reduce distress and enable coping mechanisms which may help minimize clinical symptoms as a consequence of a disturbing test result in the aftermath of PND. Our data showed that those who were content with the decision made – no matter whether they decided to continue with their pregnancy or terminate it – showed less signs of depression and anxiety a half a year after having made the decision In other words, if we know how satisfied a woman is with the decision she made we may be able to anticipate how likely she is to develop clinical symptoms at a much later time-point – namely after she has acted upon her decision. One consequence of this should be that within the counselling procedure in PND special attention should be paid to the information given prior, during and after PND, as the better informed women feel the less clinical symptoms they will develop. Furthermore, our presumption that an empathetic and sensitive health professional will put the woman's mind at ease with the decision she has made thus relieving her distress proved to be a valid one (for further details see Fischmann in print). This particular way of care giving may help women to digest their experience better and prevent them from developing depressive or anxious symptoms.

Thus, picking up an actual political debate on late abortion in Germany, our results strongly indicate that counseling and providing relevant information before, during and after PND is pivotal. Most political parties in Germany agree on this and have unanimously agreed that prenatal counseling procedures have to be improved. What is still controversial is the question of a compulsory 3-day time period for consideration between diagnosis and action. Our questionnaire-data indicate that women were not engaged in many discussions prior to testing nor in contemplating the possible actions to be taken after a positive test result, ultimately leading to a quick decision. The majority also reported that they felt they shared the responsibility for their decision. At the same time the better informed they felt the less depressed and anxious they were in the aftermath of PND, as they would accept more the decision they made. These finding strongly support psychosocial and genetic counseling before and during PND as it may help to prevent clinical symptoms in the aftermath. It does not necessarily indicate a compulsory 3-day waiting time. Interpreting our clinical material from interviews with women after they made a decision suggests a different view. Here, women reported that they felt left alone with their decision and we often found that they reported that they had quickly jumped to a decision. Time to contemplate that decision would seem necessary psychotherapeutically. Going into action is a form of defence that erroneously promises quick relief; in the aftermath though this leads to many conflicts if the decision made was a wise one.

From these findings, I would strongly advocate the development of counselling guidelines for health professionals. These should combine the expertise of different disciplines. I would also advocate a 3-day time of consideration before acting on the decision made, but would not necessarily make it compulsory as this would put much pressure on the decision-making process. A good example for counseling procedures is already realized in the Integrated Genetic Counselling Model in Italy. This interdisciplinary onsite liaison-model consisting of gynecologists, geneticists and psychotherapists with regular clinical meetings could be easily implemented by coordinating the existing health care professionals and counsellors. As a result of the EDIG study, such a liaison-model has been initiated in Germany between Prof. Dr. E. Merz and his team of gynaecologists at the Hospital Nordwest and psychoanalysts of the Sigmund-Freud-Institute led by Prof. M. Leuzinger-Bohleber. Relying on our clinical experience within the study, the finding of the clinical interviews clearly showed that an empathic and professional dialogue between the medical staff conveying a problematic finding and women/couples is important for the short-term and may support long-term coping. The knowledge gained in the EDIG study can help professionals to be 'good real objects' to women in the traumatic situation after positive findings in PND. The list of protective and risk factors – one result of the study – could be a starting point for training medical staff by raising their awareness of possible outcomes enabling them to refer patients to psychoanalysts when indicated. Once this is established a liaison between medical staff and counselors/therapists should be implemented with regular clinical conferences. Within such a liaison, the trained doctor is provided with feedback preparing him to communicate the diagnosis in a sensitive way, acknowledging his medical competence while providing him with resources to cope with the patient's grief. He can refer the patient to psychological support if he finds this necessary, based on his knowledge of risk factors. Such a liaison model will help the patient to encounter a containing and holding atmosphere relieving her of pain as she can separate and project her unconscious emotions onto the care givers enabling the patient in the process to recover ejected emotions and integrate them to the self in a more complete manner.

References

Alexander GR (1998) Preterm birth: etiology, mechanisms and prevention. Prenat Neonatal Med 3:3–9

Baldus M (2006) Von der Diagnose zur Entscheidung: Eine Analyse von Entscheidungsprozessen für das Austragen der Schwangerschaft nach der pränatalen Diagnose Down-Syndrom. Klinkhardt, Forschung, Bad Heilbrunn

Fischmann, T (2010) Ethical dilemmas resulting from modern biotechnology – between Skylla and Charybdis. Unpublished doctoral habilitation, University of Kassel

Green JM, Hewison J, Bekker HL, Bryant LD, Cuckle HS (2004) Psychosocial aspects of genetic screening of pregnant women and newborns: a systematic review. Health Technol Assess 8(33):1–109

Honnefelder L (2000) Screening in der Schwangerschaft: Ethische Aspekte. Dtsch Ärztebl 97(9):A-529–A-531

Iles S, Gath D (1993) Psychiatric outcome of termination of pregnancy for fetal abnormality. Psychol Med 23(2):407–413

Leuzinger-Bohleber M, Engels E-M, Tsiantis J (eds.) (2008) The janus face of prenatal diagnostics – a European study bridging ethics, psychoanalysis and medicine. Karnac Books, London

Salvesen KA, Oyen L, Schmidt N, Malt UF, Eik-Nes SH (1997) Comparison of long-term psychological responses of women after pregnancy termination due to fetal anomalies and after perinatal loss. Ultrasound Obstet Gynecol 9(2):80–85

Zeanah CH, Dailey JV, Rosenblatt MJ, Saller DN (1993) Do women grieve after terminating pregnancies because of fetal anomalies? A controlled investigation. Obstet Genecol 82:270–275

Zigmond AS, Snaith R (1983) The hospital anxiety and depression scale. Acta Psychiatr Scand 67(6):361–370

Chapter 4
Reconstruction of Pregnant Women's Subjective Attitudes Towards Prenatal Diagnostics – A Qualitative Analysis of Open Questions

Katrin Luise Läzer

Abstract Over the last two decades, enormous progress in prenatal diagnostic techniques has enabled us to visualize human life before birth. This would previously have been unknown to the mother and may affect the way in which she processes the pregnancy psychically. This chapter deals with the "qualitative content analysis" (Mayring, Qualitative Inhaltsanalyse, Grundlagen und Techniken, Deutscher Studien Verlag, Weinheim, 1988; Mayring, Qualitative Inhaltsanalyse, in Flick et al. (eds) Qualitative Forschung, Ein Handbuch, Rowohlt Taschenbuch Verlag, Reinbek bei Hamburg, 2000) of three open questions answered by 336 German women in the EDIG study. The first two questions should access the women's motives and feelings: "Do you have thoughts, fantasies and wishes about your developing baby?", "Have you talked to anyone about what you would do if an abnormality was detected? What did you talk about?" The third question was administered after the women had received the negative test result: "Could you describe to us how you feel now?"

Keywords Prenatal genetic diagnostics • Psychological distress • Qualitative content analysis

4.1 Introduction

Over the last two decades, enormous progress in prenatal diagnostic techniques has enabled us to visualize human life before birth and to predict some aspects of the life, health and disabilities of the baby. On the one hand this creates medically sophisticated knowledge and expertise, on the other hand pregnant women are laypersons without such expertise. Without technical support the pregnant woman's perception of her unborn baby is limited to introspective feelings, experiences,

---

K.L. Läzer (✉)
Sigmund-Freud-Institut, Myliusstraße 20, Frankfurt 60323, Germany
e-mail: laezer@sigmund-freud-institut.de

observations and her own fantasies. She – herself – doesn't know what is going on in her womb, whereas prenatal diagnostics visualizes what otherwise would remain unknown to her. These new technical possibilities might affect the way pregnancy is experienced psychically.

Studies during the past 20 years on the psycho-emotional effects of prenatal diagnostics almost unanimously show the distressing effect these tests have on women (cf. Cox et al. 1987; Sjögren and Uddenberg 1989). Regardless of the type of testing done these studies showed that prenatal diagnosis causes distress for pregnant women during the testing procedure (Sjögren and Uddenberg 1989) or has a negative emotional impact (Pauli et al. 1990). More recent studies (Kukulu et al. 2006; Leithner et al. 2008) confirm that prenatal diagnostics increases anxiety and distress and can even distinguish between high-risk patients who are more prone to psychic distress or who have a higher probability of an abnormal test result, and patients who don't show any of these risk factors. Leithner et al. (2008) as well as Kukulu et al. (2006) and Cederholm et al. (2001) strongly suggest that counselling should be offered to high-risk patients and families. Kowalcek et al. (2001) studied 324 pregnant women before and after prenatal testing and concluded that prenatal diagnostics is accompanied by psychic distress and can lead to dramatic decisional conflicts – regardless of the type of prenatal diagnostics, invasive or non-invasive testing.

These findings underline the fact that pregnant women do experience emotional distress before and after prenatal testing. Therefore it is important to find out more about the perspectives of the women and their motivations. This chapter deals with the following questions: How is prenatal diagnostics represented mentally and emotionally? What do pregnant women think and feel about prenatal diagnostics? Why do they make use of this technology? This paper tries to describe prenatal diagnostics from the point of view of pregnant women and by doing so tries to contribute to a deeper understanding of the psychological impact of prenatal diagnostics.

4.2 Method

The investigation presented in this paper is part of the European study "Ethical dilemmas due to prenatal and genetic diagnostics" (EDIG) (cf. Chap. 3; Fischmann et al. 2008; Leuzinger-Bohleber et al. 2008). One aspect of interest was women's attitudes towards prenatal diagnostics. The findings of the standardized questions and items exploring women's attitudes are presented in Chap. 3. In addition to standardized questions, *open questions* were included in the questionnaires, and it is the findings from these questions that will be presented here. But why did we use open question and what kind of method was applied?

Wanting to find out more about the individual meanings and fantasies of women undergoing prenatal diagnostics we used the questionnaire data of the EDIG study to investigate the inner psychic world of the study participants. In order to tap the partly unconscious fantasies it seemed to be more appropriate to ask open questions instead of standardized scales which predetermine the possible range of given answers (e.g.: I strongly disagree=0, I disagree=1, Neutral=2, I agree=3, I strongly agree=4).

Thoughts and feelings that go beyond that format cannot be expressed. Open questions are characterized by their wider scope, and they do not impose such limitations. Open questions provide enough space to the participants to state their own observations, thoughts and feelings. Therefore answers to open questions may contain more information and more diverse meanings than standardized questions. On the one hand the researcher has the opportunity to learn more about the inner psychic world of each participants; however, the researcher has to interpret and represent the comments made by the individuals.

The analysis presented here comprises the following steps. First, the researcher has to identify qualitative categories, which integrate and condense the answers in such a way that the content is still represented. The next step is to limit the number of categories, to enable the researcher to quantify the distribution of answers.

A bottom-up method, called "qualitative content analysis" developed by Phillip Mayring (1988, 2000), enables the researcher to generate categories out of the given answers to open questions, while allowing quantification of such categorized data. In order to ensure objectivity and reliability of the procedure a systematic approach and the validation of the derived categories are crucial. When analyzing the open questions we adhered to Mayring's recommendations:

Material was digitized keeping it anonymous. Often the given answers comprised only short remarks allowing for different interpretations. As researchers we agreed to analyse only the manifest meaning (and not the latent meaning). Based on these considerations we, a small research group of four junior scientists, coded the answers by systematically comparing and grouping the statements. This classification procedure led to qualitative categories which were controlled and, where necessary, adapted by an expert group (including amongst others Tamara Fischmann and Nicole Pfenning-Meerkötter). Next, all given answers were classified into these categories (validation of categories by an expert group – control method 1).

Secondly we chose illustrative and typical examples out of the wide range of answers for each category. The categories with their assigned references were handed over to two independent raters who used them to allocate the given answers to the respective categories. Based on their allocations we calculated the distribution of given answers to the different categories (in percentages) and the rater's agreement (Cohen's Kappa; cf. Cohen 1960; Wirtz and Caspar 2002; control method 2).

This chapter presents an analysis of responses to three questions. The first two encourage the women to tell us about their motives and feelings, namely:

1. Do you have thoughts, fantasies and wishes about your developing baby?
2. Have you talked to anyone about what you would do if an abnormality was detected? What did you talk about?

These questions were asked after women had undergone amniocentesis and were waiting for the test results. The first was an attempt to identify the main concerns of the women. The second addresses the potential problems and conflicts that could arise when confronted with a positive result after amniocentesis.

The third question was asked after women had received their test result:

3. Could you describe how you feel now?

This question is focusing on direct feelings and reactions after receiving the test result.

4.3 Findings

Three hundred and thirty six German women completed a questionnaire while waiting for amniocentesis results and almost all women (98.3%) had an amniocentesis. Two hundred and nineteen women answered the first question, and 230 the second. The third question was answered by 274 women after test results were available.

The findings are presented in tables which show the categories, examples of typical comments coded in each category and the number and % of women who gave answers coded within each category.

4.3.1 First Question: "Do You Have Thoughts, Fantasies and Wishes About Your Developing Baby?"

All but eight women (96.4%) expressed the wish that their developing baby would be healthy and develop normally (categories 1, 2, 3, Table 4.1). Their wish for a healthy child was the most important consideration.

Differences in responses were in terms of *how* the issue of health and well-being of the child was discussed. Category 1 summarized the statements of those 68% of women, who gave words like "healthy", "health" and "normal" or "normal development". Typical comments were: *"I wish for a healthy child for our family"*, or: *"that my child will be born healthy"*.

The second group comprised those 25.7% of women, who in addition expressed the wish that their baby should have certain traits or attitudes. They not only wished for a healthy and normally developed child, but they also mentioned aspects like "fortune" and "contentment" or "happiness" in the child's life. One woman wrote: *"I wish 'our' baby good luck, happiness and health and that it keeps developing normally in a quiet and relaxed atmosphere with happy 'parents'!"* Moreover, traits and characteristics are fantasized and could be understood in term of a strong and positive bonding with the child. One woman answered the first question regarding fantasies in the following way: *"Yes, of course I have. I hope that it develops 'normally' in my belly, will be born healthy and that it develops 'normally' and turns into an intelligent and lovely human being"*.

Table 4.1 Responses to question 1: "*Do you have thoughts, fantasies and wishes about your developing baby?*"

Cat	Categories	Typical examples	Frequency N	Frequency in %
1	Addressing the health status and normal development of the child	"only healthy", "Health!", "that my baby will be born healthy", "that it develops normally", "a healthy child"	149	68.0
2	Addressing wishes concerning the child's well-being, showing a strong and positive bonding with the child	"that it becomes healthy, intelligent and very beautiful", "I am thinking about him very often, almost all the time…", "I wish for a child, that has no serious illnesses. I want him to have a happy childhood and a lot of time to develop".	56	25.7
3	Addressing anxieties and concerns that show a strong and positive bonding with the child	"that my bad health condition does not harm my baby"	6	2.7
4	Addressing anxieties and concerns that are more linked to the mother, her partner or experiences	"that my partner and I get along in the new situation", "I am very scared, because I already had a miscarriage. It's hard to get involved with my baby"	4	1.8
5	Residual category	If the answers could not be allocated to any of the categories 1–4	4	1.8
	Valid data		219	100

Rater agreement for two raters: Cohen's Kappa = 0.55

Six women expressed concerns about the health of their child because of their own poor health condition (category 3). One example was: "*Of course I hope that it is healthy and that my bad health condition does not harm my baby*".

Four other women had previously experienced traumatic events such as miscarriages and these women expressed concerns related to these experiences, for example: "*I wish for a healthy child. I am very scared, because I already had a miscarriage. It's hard to get involved with my baby*".

In summary, the findings of the first question are that all women reported concerns about the health of their unborn baby. It is however important to bear in mind that the question was posed in the context of a study of prenatal diagnosis. For that reason one could assume that the wish for a healthy child is an essential motive for women to undergo prenatal testing.

4.3.2 Second Question: "Have You Talked to Anyone About What You Would Do If an Abnormality Was Detected? What Did You Talk About?"

Two hundred and thirty women, 60% of those who completed the questionnaire, answered this question. Based on a satisfactory rater agreement (Cohen's Kappa = 0.69) Table 4.2 shows the following five categories of response:

Thirty-six women preferred not to think about the possibility of an abnormal test result or they were worrying about it all the time but still couldn't make a decision

Table 4.2 Responses to question 2: "*Have you talked to anyone about what you would do if an abnormality was detected? What did you talk about?*"

Cat	Categories	Typical examples	Frequency N	Frequency in %
1	Addressing the possibility of termination of pregnancy *independent* of the grade of disability	"I thought about whether I should have the baby at all!", "We decided to terminate the pregnancy"	126	54.8
2	Addressing the possibility of termination of pregnancy *dependent* of the degree of disability	"at what grade of disability and what degree of probability I would terminate the pregnancy", "not to have the child, if it was seriously disabled"	21	9.1
3	Indecisiveness, imprecise statement, repressing the test result with a tendency towards optimism	"We haven't decided yet", "what if", "we decide, when we will have the test result", "I very much believe that everything will be ok"	36	15.7
4	Addressing the future life and possible consequences, anxieties and uncertainties	"I asked other parents for their opinions and experiences", "Who will care for the child, when I am too old. I would be forced to place it in a foster home", "uncertain future for the child and the family"	41	17.8
5	Addressing the possibility of continuing pregnancy	"We would accept a disabled child"	4	1.7
6	Residual category	If the answers could not be allocated to any of the categories 1–5	2	0.9
	Valid data		230	100

Rater agreement for two raters: Cohen's Kappa = 0.69

(15.7%, category 3). They postponed their decision to a later time point (e.g. *"We haven't decided yet"*) or their statements expressed no opinion (e.g. *"what if"*). Several women adopted a very optimistic attitude, like the following woman: *"I very much believe that everything will be ok"*. All women in category 3 seemed to share a tendency towards avoiding a possible conflict by adopting a very optimistic stance or by refusing further considerations.

The majority of the women (83.4%) raised the issue of possible disabilities and anticipated that a decision might have to be made (Categories 1, 2, 4, and 5). The possible decisions, as well as the decisional processes, varied among the pregnant women:

One hundred and twenty six women (54.8%, category 1) reported that they would consider a termination if a disability was detected with a typical comment being: *"If an abnormality was detected, a termination of pregnancy would take place"*. This group of women, although speaking hypothetically, seemed to be very certain of their decision.

A second group of women (9.1%, category 2) reported a clear tendency towards a termination of pregnancy, but gave a more differentiated view on possible disabilities. In their view the decision towards a termination would be based on the severity or type of the disability, for example: *"We will not have a seriously disabled child"* or: *"termination if mentally disabled. Otherwise we would not have done the test"*. Here, the decisional process seemed to be more differentiated. Overall, 147 women (63.9%) were predisposed towards termination in the case of an abnormal test result, a similar proportion to those who responded to the direct question – that they would *"terminate a pregnancy if there is something wrong with the baby"*. (cf. Chap. 3.)

A fourth group of women reported their concerns and fears about having to make a decision (17.8%, category 4). They reported thoughts about their future life with a disabled child. For example, one woman reported: *"I talked to a pregnant colleague. A serious defect was found when she was 13 weeks pregnant. She helped me a lot with her experiences"*. Another wrote: *"Birth yes/no?; proper care for disabled child?; future prospects for the child, social integration; change of the personal environment of the parents"*. Other women reported concern about the impact on their family life and existing children: *"That I am responsible for the unborn child as well as for my first son"*. *"What does it mean to have a disabled child? Who will care for him? What can we offer him? Am I egoistic? What is the effect for partnership/family?"* These 41 women gave answers that appeared to show a decision process that was still open and reflective about prenatal testing, with thoughts about future anxieties identified.

Four women (1.7%, category 5) reported that they would consider or had decided to continue the pregnancy. Prenatal diagnostics are perceived as a means of preparing for the birth of the baby.

4.3.3 Third Question (After Receiving the Negative Test result): "Could You Describe How You Feel Now?"

The third question reported here was asked of women after they had received their test result and all women (N = 274) expressed their relief after having received the negative test result. (This analysis (cf. Table 4.3) only includes data from women who received a negative finding).

Most women (94.9%) are relieved and happy (categories 1 and 2). The majority expressed their relief and happiness in a direct manner. A typical answer was: *"Good, happy and relieved."* Most women did not mention their feeling during the waiting phase. What one could conclude is that all anxieties and psychic strains are merely implicitly expressed by using terms such as "relieved" and "happy".

Table 4.3 Categories and frequencies with respect to question 3: *"Could you describe how you feel now?"*

Cat	Categories	Typical examples	Frequency N	Frequency in %
1	Happy and relieved	"Very relieved and happy about the result", "Relieved that my child is healthy! I am very happy and can prepare for the birth, because now I know the sex.", "I have never thought it could be a negative finding, nevertheless the test makes me feel secure."	194	70.8
2	"Now let's start off!" "Now I look forward to the baby" after having gone through days of sorrow and anxiety	"Big relief, now I really feel pregnant.", "Happy, happy, happy. Great, it is a fantastic feeling. I am totally relieved. I've gone through weeks of sorrow and anxiety."	66	24.1
3	Relieved and happy, however, feelings of uncertainty, disappointment, disillusionment remain to some degree	"Very happy because the result is ok. Naturally, some uncertainty remains as the whole pregnancy up to birth is risky and very exciting"	13	4.7
4	Residual category	If the answers could not be allocated to any of the categories 1–3	1	0.4
	Valid data		274	100

Rater agreement for two raters: Cohen's Kappa = 0.73

A similar tendency might apply to women of category 2 (24.1%) who conveyed their uncertainty and anxieties by accentuating the term "now". "Now" let's start off. "Now" they are looking forward to the baby. Some examples: "*Big relief, now I really feel pregnant*", "*Happy, happy, happy. Great, it is a fantastic feeling. I am totally relieved. I've gone through weeks of sorrow and anxiety*" (cf. Katz Rothman 1986; Chap. 5).

A third group of women (category 3; 4.7%) expressed feelings of relief, while still reporting feelings of uncertainty. A typical answer was: "*Yesterday I felt relieved and relaxed. Today I think that 'by accident' my cells were analyzed instead of the child's cells. I have to talk to my doctor about this and the probability of such a mistake.*" The test result could not reassure her. Thus, although relief and happiness were the dominant response, a few women continued to worry about the pregnancy and the health of the unborn baby.

Concerning psychic distress, these results could be summarized and illustrated as following. Different studies have shown that prenatal diagnostics leads to increased levels of distress (cf. Cox et al. 1989; Sjögren and Uddenberg 1989; Pauli et al. 1990; Cederholm et al. 2001; Kowalcek et al. 2001; Kukulu et al. 2006; Leithner et al. 2008). Our data suggests that some women experience very intensive feelings of distress and worry connected to the testing procedure until they received the negative result. To give an example: "*I feel very relieved; the enormous tensions – which in retrospect had developed unconsciously and became conscious only after having learned about the result – fell off my mind and my body*." Several times women referred to a German idiom: "*Ein dicker Stein fällt vom Herzen*" – "A heavy stone falls from the heart" which can be translated as: "It took a load off my mind". By this metaphor, the women expressed how burdened they felt during the phase in which they waited for the results. Some women even started to cry when they received the test result – a psycho-physiological expression of relief. One woman directly addressed the connection between her distress and her tears: "*I am relieved and very happy, however, the last weeks of distress also made me feel exhausted and worn out. When I read the letter, I cried; I am still close to tears.*"

Some women uttered their gratitude. Two examples: "*Simply relieved and indeed very glad and grateful!*" or: "*I cried for joy. I am so happy. Incredible. I thank God with all my heart.*"

All quotations can be read as reference to the distress the women experienced in the waiting phase after an amniocentesis.

These presented findings and quotations underline that pregnant women might experience intensive and idiosyncratic feelings of distress during the prenatal procedure. Therefore medical health professionals should acknowledge the possibility of the emergence of such feelings of distress.

References

Cederholm M, Sjödén P-O, Axelsson O (2001) Psychological distress before and after prenatal invasive karyotyping. Acta Obstet Gynecol Scand 80:539–545

Cohen J (1960) A coefficient of agreement for nominal scales. Educ Psychol Bull 10(1):37–46

Cox DN, Wittmann BK, Hess M, Ross AG, Lind J, Lindahl S (1987) The psychological impact of diagnostic ultrasound. Obstet Gynecol 70:673–676

Fischmann T, Pfenning N, Läzer KL, Rüger B, Tzivoni Y, Vassilopoulou V, Ladopoulou K, Bianchi I, Fiandaca D, Cirillo I, Sarchi F (2008) Empirical data evaluation on EDIG (Ethical dilemmas due to prenatal and genetic diagnostics). In: Leuzinger-Bohleber M, Engels E-M, Tsiantis J (eds) The janus face of prenatal diagnostics: a European study bridging ethics, psychoanalysis and medicine. Karnac, London, pp 89–135

Giddens A (1992) Konsequenzen der Moderne. Suhrkamp, Frankfurt

Katz Rothman B (1986) The tentative pregnancy: how amniocentesis changes the experience of motherhood. W.W. Norton, New York

Kowalcek I, Mühlhoff A, Bieniakiewiz I, Gembruch U (2001) Nicht-invasive und invasive Pränataldiagnostik und psychische Beanspruchung der Schwangeren. Geburtsh Frauenheilk 61:593–598

Kukulu K, Buldukoglu K, Keser I, Simsek M, Mendilcioglu I, Lüleci G (2006) Psychological effects of amniocentesis on women and their spouses: importance of the testing period and genetic counselling. J Psychosom Obstet Gynecol 27:9–15

Läzer KL (2007) Einstellungen schwangerer Frauen zur pränatalen und genetischen Diagnostik. Kategorisierung, Quantifizierung und Auswertung offener Fragen. Unveröffentlichte Diplomarbeit, Freie Universität, Berlin

Leithner K, Pörnbacher S, Assem-Hilger E, Krampl E, Ponocny-Seliger E, Prayer D (2008) Psychological reactions in women undergoing fetal magnetic resonance imaging. Obstet Gynecol 111:396–402

Leuzinger-Bohleber M, Engels E-M, Tsiantis J (eds) (2008) The janus face of prenatal diagnostics: a European study bridging ethics, psychoanalysis and medicine. Karnac, London

Mayring P (1988) Qualitative Inhaltsanalyse. Grundlagen und Techniken. Deutscher Studien, Weinheim

Mayring P (2000) Qualitative Inhaltsanalyse. In: Flick U, von Kardorff E, Steinke I (eds) Qualitative Forschung. Ein Handbuch. Rowohlt Taschenbuch, Reinbek bei Hamburg

Pauli Ch, Blaser A, Herrman U (1990) Amniozentese: Psychische Belastung und deren Bewältigung bei der schwangeren Frau. Geburtsh Frauenheilk 50:291–294

Sjögren B, Uddenberg N (1989) Prenatal diagnosis and psychological distress: amniocentesis or chorionic villus biopsy? Prenat Diagn 9(7):477–487

Wirtz M, Caspar F (2002) Beurteilerübereinstimmung und Beurteilerreliabilität. Hogrefe. Verlag für Psychologie, Göttingen/Bern/Toronto/Seattle

Chapter 5
Prenatal Testing: Women's Experiences in Case of a Conspicuous Test Result

Elisabeth Hildt

Abstract This chapter will focus on ethical aspects and decision-making conflicts in prenatal testing that arise in the case of conspicuous test results. In order to gain some insight into the women's experiences when confronted with a conspicuous test result, we studied their answers to the open questions in the questionnaires of EDIG substudy A. After giving a short introduction to the open questions some of the responses will be presented. The focus will be on an aspect which plays a crucial role in the overall experience: the women's attitude towards the fetus – and the way women talk about the fetus. The results of the EDIG study will then be discussed against the background of Barbara Katz Rothman's concept of a "tentative pregnancy". Afterwards, some implications for the ethics debate will be discussed.

Keywords Ethics • Prenatal testing • Qualitative research • Tentative pregnancy • Termination of pregnancy

Prenatal testing is a multifaceted biomedical technology. For pregnant women there is the option to undergo prenatal testing in order to find out whether the fetus carries a certain abnormality. In most cases, prenatal testing serves to reassure pregnant women with an inconspicuous test result indicating that no genetic or structural variation has been found in the fetus. The various methods of prenatal testing can be considered, on the one hand, as medical procedures which offer valuable options based on the test result. On the other hand, however, in prenatal testing, pregnant women are confronted with decision-making conflicts which are particularly serious when a fetal abnormality is detected. In the case of a conspicuous test result, indicating a serious non-treatable fetal condition, women and their partners have the option to decide on whether to continue or terminate the pregnancy. These are very complex and onerous decision-making conflicts for the individuals involved,

E. Hildt (✉)
Department of Philosophy, University of Mainz, Jakob-Welder-Weg 18,
55099 Mainz, Germany
e-mail: hildt@uni-mainz.de

raising manifold practical and ethical issues (Statham 2002; Hildt 2002, 2008; Leuzinger-Bohleber et al. 2008).

In the European research project "Ethical Dilemmas due to Prenatal and Genetic Diagnostics" (EDIG), questionnaires and interviews have been carried out in order to explore the experiences and decision-making conflicts of couples during prenatal diagnosis and the associated ethical dilemmas (cf. Leuzinger-Bohleber et al. 2008). The emphasis of the following contribution is on the situation experienced by women who received a conspicuous test result. Women's answers to open questions in the EDIG study questionnaires of substudy A are particularly helpful in this respect for they directly reflect their experiences and decision-making conflicts.

In what follows, after a short introduction to the EDIG questionnaires, some of the qualitative results of the EDIG substudy A will be presented and discussed. The focus will be on the women's attitude towards the fetus, in particular on the way women talk about the pregnancy and the fetus.

5.1 The EDIG Questionnaires: Open Questions

In the EDIG substudy A, questionnaire studies have been carried out in Greece, Israel, Italy, and Germany (cf. Fischmann et al. 2008; Chap. 3 in this volume). Pregnant women who attended a medical doctor or a clinic for prenatal testing were asked to participate, i.e. to fill in several questionnaires at several time points: when they first visited the medical doctors for prenatal testing (screening questionnaire), when they waited for the test results (waiting period questionnaire, WaitP), within 3 weeks after testing (decision phase questionnaire, DeciP), and 6 weeks after testing (digestion phase questionnaire, DigeP). In addition, there was a follow-up-questionnaire (Foll) which was sent about 8 months after the test result, i.e. after the end of the pregnancy. The questionnaires comprised open and closed questions and used standardized instruments. In the EDIG study only those women who underwent prenatal testing participated. The EDIG study does not cover the pre-testing phase, nor does it cover the views of women who chose not to use prenatal testing.

In what follows, the focus will be on the answers that women who received a conspicuous test result gave to the open questions. Typical open questions are: "Could you describe how you feel now?" (DeciP); "What are the factors you are considering when making your decision?" (DeciP); or "Is there anything you would like to say about the way in which your own experience has affected your views on prenatal diagnosis?" (DigeP+Foll). In addition, there were several open questions on professional support (DigeP+Foll). So, with regard to the open questions the women were free to write their own answers and by this to express their own ideas.

The analysis carried out here is confined to the data obtained in Germany. Data has been obtained from women and their partners who visited gynaecologists and clinics in Frankfurt, Mainz and Offenbach. In Germany, 864 participants (women and their partners) were recruited. Most of the women underwent amniocentesis. Whereas most participants received an inconspicuous test result, there were 22 women who got a conspicuous test result. Fourteen of these women responded in detail to the open questions and their answers are taken into consideration in the following discussion. Four women continued their pregnancy, whereas ten decided to terminate their pregnancy. Terminations were carried out following the discovery of conditions such as Trisomy 21, Trisomy 18, Turner's syndrome, or multiple organ abnormalities, detected by ultrasound scanning. Among the four women who continued their pregnancy, two women continued with a Down syndrome baby, one woman continued after having received the result "teratoma in the tailbone" ("Steißbeinteratom"), another after having received the test result "Hygroma colli" (accumulation of fluid in a cyst).

It is important to stress that the analysis carried out here based on the open EDIG questions is purely qualitative research. The answers comprise those considerations which women wrote on their own initiative. This does not indicate that the aspects they gave are the only ones that are relevant to them, nor does it indicate that those women who did not give an answer to the questions did not engage in intensive reflections on prenatal testing. Thus, the answers merely hint at the aspects considered important by the women; they are in no way exhaustive.

In addition, the data consists of the subjective views of a small group of women who participated in the EDIG study in Germany and who gave their comments to the open questions. The results are in no way representative. This is particularly true for the very small group of women who received conspicuous test results. Furthermore, there is a bias insofar as there is a special subgroup of women who – for one reason or another – are motivated to comment at length on the questions. Those women for whom the issues raised in the questionnaires are not of so much interest, those who feel embarrassed or those for whom the decision-making conflict is extremely disturbing most probably did not answer the open questions. Needless to say we can only discuss the answers we actually received. That is why those women who gave detailed answers are considered in a disproportionate way. In spite of these limitations, however, the answers obtained are of enormous relevance to us since they help us draw a picture on the experiences of the women involved.

5.2 How Do Women Who Received a Conspicuous Test Result Talk About the Pregnancy and the Fetus?

When a conspicuous test result indicates serious fetal health problems a very problematic situation arises: women have to decide on the life of the fetus. They have to decide on the future course of their pregnancy which was, – most probably, – highly wanted before receiving the test results. How do these women perceive the fetus? How do they talk about the fetus?

There is no uniform view. Instead, women clearly differ in their answers. Some women may perceive the fetus in a rather neutral way as is suggested by the following answer to the question: "What have you been told about the options that are available to you?"[1]

---

[1] Answers translated by E. Hildt.

> I can get rid of it [the baby], I can keep it or I can give it up for adoption. (DeciP)

Other answers point to a quite intense relationship with the fetus as can be inferred from the way women write about the fetus and the overall situation. For example, when talking about the termination of her pregnancy, one woman writes

> On Monday 16.4. the amniocentesis was carried out, on Wednesday 18.4. I received the test result, on Friday 20.4. our child was born. (DeciP)

Sentences like these suggest the birth of a child and not the termination of a pregnancy. Only the context in which the woman describes the situation indicates that in this case, there has not been a normal birth, but the termination of the pregnancy. The woman talks about the fact that her *child* was born, she does not talk about a fetus, and does not mention that the child will not live. The perspective the woman takes is one on her (future) child, and not a rather neutral view of a fetus.

In addition, some women have given a name to their fetus, such as "Liliput" or "Lis", and by this talk about their fetus or future child in a personified way. The phenomenon of naming can be seen from the waiting phase questionnaire onwards.

It seems that at least some women have established some – albeit partial – sort of interpersonal relationship with the fetus, as is suggested by the following:

> I underwent amniocentesis simply because for some weeks I had problems with my partner. I couldn't imagine bringing up a disabled child as well as my 5-year-old son all by myself. And contrary to my expectations I promptly received a Trisomy 21 result. Inconceivable! Disaster! A black day! I am sorry for my little Lis but I have to go through with the abortion. I hope she will forgive me! (WaitP)

The woman reports that she had an amniocentesis only because of problems with her partner, and that she unexpectedly received a Trisomy 21 test result. Then she talks about little "Lis" and her plan to have an abortion. She hopes "Lis" will forgive her. In these sentences the woman not only expresses feelings of guilt, of being sorry for what she will do to the fetus, but she also enters a kind of moral discourse with her fetus, her little "Lis". In this, she assumes the fetus to be a person who – hopefully – will forgive her for what she has done to her.

Another woman writes in the digestion phase questionnaire about her decision to terminate the pregnancy after having received a Trisomy 18 test result:

> The mind still says that the decision was the right one for all of us. Nevertheless sometimes the question arises what would have happened if we had decided to let our little one decide about his time of death… (DigeP)

She says that she knows that the decision to terminate the pregnancy was right. However, instead of talking about the situation in an impersonal, neutral manner as may be expressed by phrases like "let nature take its course" the woman sees the fetus in a personified way. The idea to let "our little one" take the decision on when to die implies the context of informed consent and surrogate decision-making and reflects the idea that it might have been better to let the person involved decide on his own when to die. So, the woman does not consider the fetus to be a mere object, but she considers the fetus to be a person who – in a certain way – is able to take decisions.

In the follow-up questionnaire, the same woman gives an account of the funeral that has taken place after the pregnancy termination:

> We buried our Joshua in a nice wooded cemetery nearby on a burial ground for stillborn children up to 2,000 g. Only my husband, our 'eldest' (3 ½ years old) and a hospital clergyman attended. The clergyman found very fair and comforting words for us, so we have good memories of the burial. A small granite plate with the name decorates the little grave. We are glad to have the grave as a place of refuge. (Foll)

She speaks about the religious committal service they had for Joshua, the grave and the gravestone; she also says she is glad to have a place of refuge, the grave. Religious committal services, a grave with the name written on a gravestone – all of these are clear indications of the loss of a beloved human being.

The answers cited above indicate that some of the women do consider – at least implicitly – the fetus to be some kind of person or moral agent. They have established some – albeit partial – form of mother-child-relationship during the pregnancy. It is not a neutral biological entity they decide on, but a person they enter into moral discourse with. In view of this individual relationship, in the case of a conspicuous test result, the decision-making conflict is an extremely problematic and ambivalent situation. For the women, to decide on life and death of their fetus, their future baby they got intensely involved with, is a very hard decision.

The ambivalence of the situation is directly expressed by the following sentences, where the woman complains about her doctor:

> Personally, I did not like a statement made by our gynaecologist (directly after the amniocentesis). She said that, while waiting for the test result, we should distance ourselves from the pregnancy, from the baby, and pretend not to be pregnant!! I told her that I will not give up my baby just like this and that in fact I plan to do exactly the opposite! (DigeP)

Whereas the doctor suggests mentally embracing the pregnancy only after receiving an inconspicuous test result, the pregnant woman stresses the reality of her ongoing pregnancy and her relationship with her fetus. While the woman talks about her baby, the doctor relies on the concept of "tentative pregnancy", assuming that it would be better to wait for the test result before preparing for the future child. The concept of "tentative pregnancy" which has been coined by Barbara Katz Rothman (1993) will be discussed in more detail.

5.3 The Concept of Tentative Pregnancy

The concept of a tentative pregnancy has been developed by Barbara Katz Rothman in her very influential book "*The Tentative Pregnancy: Prenatal Diagnosis and the Future of Motherhood*", first published in 1986. In this book she describes the influences of amniocentesis and similar technologies on the way women experience pregnancy and motherhood and she critically reflects on the future of motherhood. One of her main theses is that prenatal genetic testing encourages not only women but society as a whole to consider children to be products of conception that undergo quality control; that prenatal genetic testing enormously influences mother-child-relationships insofar as pregnant women – before receiving an inconspicuous, negative test result – tend to consider their pregnancy as a "tentative pregnancy"; a pregnancy which they fully embrace only after they have come to know that "everything is alright" (Rothman 1993).

Some results of the EDIG study also point in this direction. Within the EDIG study questionnaires, most of the participants who received an inconspicuous test result in some way or another expressed their relief about the test result. In the case of an inconspicuous test result, after a period of uncertainty and inner tension the women feel relieved by the test result. Among them, some women addressed ideas that are very similar to the concept of a "tentative pregnancy" coined by Barbara Katz Rothman. The women did this spontaneously, i.e. without having been stimulated by a specific question relating to the "tentative pregnancy" concept. For example, in the Decision Phase Questionnaire, after having received an inconspicuous test result, some of the answers to the open question "Could you describe how you feel now?" reflect this concept:

> Very, very relieved! It's a load off my mind. The gloomy feeling of the last few days is disappearing slowly, at the end I slept badly. Now I finally want to enjoy the pregnancy, get involved in the pregnancy, be happy – things about which one can read a lot and which were very distant from me up to now. (DeciP)

> [I feel] confirmed in my positive foreshadowing and relieved. Now I can tell others about the pregnancy, without reservations (such as 'if everything goes well'/'let's await the test results'). (DeciP)

> I am very relieved; before the tests I didn't allow myself to get involved with the baby, I didn't tell third parties about the pregnancy which I did directly after the testing. (DeciP)

With these sentences, the women describe the influence of prenatal genetic testing on their way of experiencing the pregnancy. They report on their emotional reservations before receiving an inconspicuous test result. Some women did not tell their family or friends about being pregnant before receiving the negative test result.

This suggests that at least some women are very reluctant to embrace the pregnancy and to tell others about their pregnancy before receiving a result that indicates that "everything is alright".

5.4 Relationship and the Tentative Pregnancy Dilemma

Before knowing the test result, women who undergo or plan to undergo prenatal testing can be considered to be a rather homogeneous group – even if of course the women differ enormously in their individual risk factors and motives for choosing prenatal testing. For they represent the group of women who opt for prenatal testing in order to get more detailed information on the fetal health status. After having received the test results, however, there are clearly two different groups: Whereas in general, women who received an inconspicuous test result feel relieved and continue their pregnancy in a reassured state, for the women who received a conspicuous test result, the situation is totally different and highly dilemmatic.

Overall, the answers the women gave to the open EDIG study questions clearly underline the ambivalence that goes along with prenatal testing. In general, for the women who choose to undergo prenatal genetic testing a strong dilemma arises which Barbara Katz Rothman has characterized as follows (Rothman 1993, p. 6):

> [The technology of amniocentesis and selective abortion] asks women to accept their pregnancies and their babies, to take care of the babies within them, and yet be willing to abort them. We ask them to think about the needs of the coming baby, to fantasize about the baby, to begin to become the mother of the baby, and yet be willing to abort the genetically damaged fetus. At the same time. For twenty to twenty-four weeks.

In some way or another, each woman who undergoes prenatal genetic testing has to face this dilemma and the resulting inner tension. On the individual level as well as on the societal level it cannot be denied that prenatal genetic testing gives support to the idea of what has been called a "tentative pregnancy".

This dilemma is clearly reflected in the above answers given by some of the women with inconspicuous test results in the EDIG study. Some report having experienced some sort of "tentative pregnancy", of having been able to fully embrace the pregnancy only after having obtained an inconspicuous test result. For the women, prenatal testing challenges the pregnancy experience in order to leave the option available to terminate pregnancy. For most women, i.e. for the vast majority of pregnant women who receive an inconspicuous test result, this period of doubt is the price for getting reassurance.

Furthermore, the EDIG results underline the intensity and the moral dimension of the decision-making conflict for the women involved in the case of a conspicuous test result. The women have to decide on the life of the fetus, after having established some – albeit partial – form of mother-child-relationship and while perceiving the fetus in a personified way. The answers given by the women with conspicuous test results indicate how women get emotionally involved with the ongoing pregnancy and consider the fetus to be their future child. For the women, prenatal testing leads to a very disturbing situation in which they are torn into two directions. On the one hand, there is the idea of "tentative pregnancy". On the other hand, it is of course strongly counterfactual to expect a pregnant woman not to identify with her ongoing pregnancy. In the case of a conspicuous test result, these basic perspectives have to be directly confronted, leading to serious conflicts when women have to decide on the course of their wanted pregnancies.

In addition, with regard to women's view on the fetus, the results point to the central relevance of relational aspects for the women involved. This is in sharp contrast to the ongoing and very controversial scientific debate on the moral status of human embryos and fetuses, which is characterized by positions relating to the right to life, human dignity, and aspects such as continuity, potentiality, identity or proportionality.

In the ethics debate on the moral status of human embryos or fetuses, the embryo or fetus is typically seen as a separate entity. This differs enormously from the context-related way in which the women involved express their experiences. In none of the answers the women gave to the open questions of the EDIG questionnaires, were these theoretical concepts and arguments ever mentioned. It seems that for the women, they are too abstract a category. Instead, for the women, relational aspects play an important role, i.e. the relationship with the fetus and future child. It is by way of these relational aspects that they reflect some of the moral issues involved.

In view of this, it is necessary to better include relational aspects in the debate on ethical issues in prenatal testing. Such a strategy aims at taking the women's point of view into consideration. This may help to more directly support the women in the enormous decision-making conflict they experience.

References

Fischmann T, Pfennig N, Läzer KL, Rüger B, Tzivoni Y, Vassilopoulou V, Ladopoulou K, Bianchi I, Fiandaca D, Sarchi F (2008) Empirical data evaluation on EDIG (ethical dilemmas due to prenatal and genetic diagnostics). In: Leuzinger-Bohleber M, Engels E-M, Tsiantis J (eds.) The janus face of prenatal diagnostics: a European study bridging ethics, psychoanalysis, and medicine. Karnac, London, pp 89–135

Hildt E (2002) Autonomy and freedom of choice in prenatal genetic diagnosis. Med Health Care Philos 5(1):65–71

Hildt E (2008) Moral dilemmas and decision-making in prenatal genetic testing. In: Leuzinger-Bohleber M, Engels E-M, Tsiantis J (eds.) The janus face of prenatal diagnostics: a European study bridging ethics, psychoanalysis, and medicine. Karnac, London, pp 273–287

Leuzinger-Bohleber M, Engels E-M, Tsiantis J (eds.) (2008) The janus face of prenatal diagnostics: a European study bridging ethics, psychoanalysis, and medicine. Karnac, London

Rothman BK (1993) The tentative pregnancy: how amniocentesis changes the experience of motherhood. Norton, New York

Statham H (2002) Prenatal diagnosis of fetal abnormality: the decision to terminate the pregnancy and the psychological consequences. Fetal Matern Med Rev 13:213–247

Chapter 6
Caring for Women During Prenatal Diagnosis: Personal Perspectives from the United Kingdom

Helen Statham and Joanie Dimavicius

Abstract This chapter reviews developments in prenatal testing. The changes in the range of tests that are available, perspectives on testing and the option of termination after a positive prenatal diagnosis are addressed. Caring for women undergoing prenatal diagnosis is a demanding task for health professionals. To enhance informed choice and support to women and their partners, health professionals must first be aware of their own values regarding sensitive issues in providing information about tests and the outcome of pregnancies after a diagnosis. The chapter details the process of testing and identifies particular times when parents have specific needs – making a decision, undergoing a termination, continuing a pregnancy. It also considers the resources necessary for professionals including their own need for training and support. Finally it identifies a number of online resources available to professionals and parents that can give further information and support.

Keywords Health professionals · Prenatal testing · Standards of care · Support groups

6.1 Introduction

The EDIG study has explored the experiences of pregnant women (and their partners) who have undergone prenatal diagnosis during 2006–2008 in a number of countries across the European Union. In this chapter we would like to reflect on our personal observations and experiences during the EDIG study. These reflections are

---

H. Statham (✉)
Centre for Family Research, University of Cambridge, Cambridge CB2 3RF, UK
e-mail: hes11@cam.ac.uk

J. Dimavicius
Concept Communications, Sextons Yard, Docking, Norfolk PE31 8NH, UK
e-mail: jdimav@aol.com

in the context of what we know takes place within the United Kingdom (UK) and our experiences, over more than 20 years, with parents undergoing prenatal diagnosis and with professionals working in the field. Our reflections are informed by the broader literature on parents' experiences of prenatal diagnosis, decision-making and living with the consequences of the decision that was made, whether to terminate the pregnancy or continue, but this chapter will focus on issues for health professionals working with women. The EDIG study data presented so far (Fischmann et al. 2008) has related to individual and personal factors and their relationship with aspects of coping with their situation, but parental experiences are also mediated by how they are cared for by the variety of health professionals involved during the process of prenatal screening and diagnosis. The chapter will begin with a brief review of developments in techniques of, and attitudes to, prenatal testing (summarized from an earlier report, cf. Statham 2002).

6.1.1 Developments in Prenatal Testing

The technological possibilities for the prenatal diagnosis of fetal abnormalities have grown enormously since the early 1900s. At that time, X-rays could detect some structural abnormalities (Case 1917, cited in Resta 1997). The discovery of the Barr body allowed human male and female cells to be distinguished (Moore et al. 1953) and by 1956 it was possible to collect amniotic fluid from pregnant women and to determine the sex of fetuses using cells from that fluid (Serr et al. 1955; Fuchs and Riis 1956; Makowski et al. 1956; Shettle 1956). This raised the possibility that women with a family history of sex-linked disorders could be offered information about their unborn baby and could potentially use that information to make reproductive decisions. A mother who was a carrier of haemophilia was reported to have undergone a therapeutic abortion following amniocentesis and the identification of a male fetus in 1960 (Riis and Fuchs 1960).

Other developments at that time included LeJeune's discovery that Down syndrome was caused when individuals had three copies of chromosome 21 (LeJeune et al. 1959) and the ability to culture amniotic cells so that the numbers of chromosomes in dividing cells could be counted. Amniocentesis became safer, by 1966 it was possible to analyse human chromosomes from amniotic fluid (Steele and Breg 1966) and some European countries began to permit legal abortion, e.g. Sweden.[1] The option of termination of pregnancies found to have abnormal chromosomes was realised in the late 1960s (Nadler 1968).

Current practice in prenatal testing, focussing on methods of screening for Down syndrome and structural malformations, methods of invasive testing and cytogenetics has been reviewed comprehensively by Bui and Meiner (2008) in an earlier publication detailing the EDIG study.

---

[1] Cf.: http://www.un.org/esa/population/publications/abortion/doc/sweden.doc (Accessed 12 April 2011).

6.1.2 Attitudes to Prenatal Testing and Abortion

Statham (2002) has previously discussed early professional attitudes to prenatal testing and the way in which it was linked with abortion. Prenatal diagnosis was initially offered to a small number of women who were at high risk of conceiving a baby with a fetal abnormality (either because of a family history of genetic disorder or a previous affected pregnancy) because only a few physicians could undertake the invasive procedure of amniocentesis safely and few laboratories could culture the amniotic samples. Given that physician time and laboratory skills were scarce valuable resources, prenatal testing was initially offered only to high-risk parents for whom '*management decisions would be changed by the data obtained*' (Evans et al. 1993), i.e. it was conditional that there was a prior commitment to abortion if an abnormality was confirmed. The way in which early positive diagnoses and terminations were managed suggested that a termination for abnormality was seen as a solution to parents' problems (Statham 1992, 1994). The baby with an abnormality was seen as a problem and the solution to that problem was to take the baby away. These attitudes impacted upon the way in which women who were undergoing prenatal diagnosis and subsequent termination were cared for. There was little understanding of parents' experiences although it was clear in even the earliest reports of parents' feelings that however much a baby with an abnormality might have been 'a problem', the decision to terminate was difficult, the experience harrowing and the psychological impact significant (reviewed in Statham 1992, 1994; Statham et al. 2000).

Over the last 40 years abortion has become available legally in most Western countries[2] and there has been a widespread introduction of a variety of prenatal diagnostic tests during antenatal care (Bui and Meiner 2008). Alongside this increasing availability of tests and the opportunities for women to undergo a termination of pregnancy, changes in the attitudes of both those offering tests and those undergoing the tests have led to different experiences. These changing attitudes derive from different perspectives in two particular areas.

First, as prenatal testing became more widely available, a new ethos was developing around genetic counselling that it should be non-directive. Decisions should be made by the individual being counselled after the counsellor has given full information (Wertz and Fletcher 1989; Kessler 1992). This view began to influence attitudes towards how decisions should be made after prenatal diagnosis, particularly among professionals haunted by fears of accusation of eugenic practices. Any link between a decision to have prenatal testing and a prior commitment to terminate the pregnancy became unacceptable (Wertz and Fletcher 1989) as it became more accepted that the decision to have a test could be no more than that, and decisions based on the result of that test could only be made once the results were available. As we will discuss below, 'non-directiveness' in counselling presents challenges, both for the counsellor, and for the woman making decisions about tests and termination. These require sensitive handling by competent professionals.

---

[2]Cf.: http://www.hsph.harvard.edu/population/abortion/abortionlaws.htm (Accessed 12 April 2011).

Secondly, attitudes towards perinatal death were changing. Bourne (1968) had suggested that stillbirths might have psychological effects on women and their doctors – until that time, women were discouraged from seeing their baby or of fostering any memories of having had a child. Kennell et al. (1970) then described the reactions of parents to the loss of a new-born infant and it was recognised that they were similar to those described by Parkes (1965) following the death of a close family member. Kennell et al. (1970) reported changes in the way they as professionals cared for parents after a baby had died, including separating bereaved parents from mothers with healthy babies and improving communication between hospital staff so that all in contact with the parents knew about the death of the baby. Once neonatal deaths were seen as bereavement, it followed for other perinatal deaths, especially stillbirths and then miscarriage. Eventually, termination for abnormality, the pregnancy loss that parents themselves choose, was recognised as a bereavement, largely as a result of the work of the parent support groups associated with this type of pregnancy loss. The different parent support groups have been important in highlighting the similarities and differences in the losses and associated distress.

Care for parents has developed, in part, based on what parents have reported that they found helpful and unhelpful. There has been little evaluation of these patterns of care which have face-validity and were introduced quickly by staff eager to ameliorate the distress they recognised in parents. Parkes (1995) has argued however that it is unethical to introduce services for the bereaved that are not well-founded and evaluated. However, whilst care has not been evaluated, UK experience within organisations such as Antenatal Results and Choices (ARC) suggests that parents now find the care they receive to be supportive.

6.2 Prenatal Testing and the Possible Outcomes

In the absence of any prenatal testing, about two percent of babies will be born with a structural or chromosomal anomaly, or a serious genetic disorder. Although a small number of women enter pregnancy at increased risk of conceiving such a baby, most anomalies occur in the pregnancies of healthy, low-risk women (RCP 1989). Prenatal testing gives women the opportunity to find out if their baby has an anomaly. *Diagnostic* tests can confirm the presence or absence of the anomaly but these tests are often expensive and come with risks, e.g. of miscarriage after amniocentesis. They will usually be offered to, or chosen by women who are at high risk when they become pregnant, e.g. because of maternal condition such as diabetes (Carrapato and Marcelino 2001; Smoak 2002; Wong et al. 2002), women who need teratogenic medication for a chronic condition such as epilepsy (Fairgrieve et al. 2000; Holmes et al. 2001; Lowe 2001) or because they have a known history of a genetic disorder in the family.

Screening tests are cheaper to perform and do not put the pregnancy at risk and are therefore used to identify from the large population of ordinary pregnant women those who are at increased risk of having a baby with an anomaly (Statham and Dimavicius 2008). It is this group of women, i.e. the vast majority of pregnant women who enter pregnancy with no reason to believe they are at increased risk of having a baby with an anomaly, who are the focus of this chapter. These women will be offered screening tests and if they decide to embark on screening they may be identified as being at high risk and are then offered a diagnostic test. It is an integral aspect of a screening test that some who have the test will be told they are low risk and do not undergo diagnosis but who will in fact have the condition being tested for (false negative) and others who are told they are high risk who find after a diagnostic test that they do not have the condition (false positive). For pregnancy screening these issues are important as we will discuss below.

It is important to remember that there are a limited number of pregnancy outcomes and that all outcomes are as likely for women who choose not to have any screening tests as well as for those who decide to have them. The majority of babies are born healthy and with no anomalies, although women may be alarmed if concerns are raised by screening tests that are subsequently shown to be false alarms. Some babies can only ever be destined to die because of the nature of the anomalies they have, e.g. anencephaly. This may be discovered at birth or during pregnancy and, if discovered as a result of tests, the baby will die whether the parents decide to terminate the pregnancy or continue: the only thing that can be changed is the time at which the baby dies. Similarly, many of the babies who are found during the pregnancy to carry non-lethal abnormalities will be born with the same life and morbidity expectancy as if they had not been prenatally diagnosed. Thus nothing changes in the care and treatment for a baby with Down syndrome whenever diagnosed as the chromosome abnormality cannot be treated; neither are there any changes in the treatment for a baby born with cleft lip and palate. Dependent upon the condition, there is surgery available and this is carried out at times related to the condition of the child and not to the time of diagnosis. For some structural anomalies, it has been assumed that prenatal knowledge might result in improved outcomes, for example if prior knowledge can result in arrangements to deliver the baby in a centre of excellence if immediate surgery is likely to be needed. However, even for conditions where improved outcomes appear a theoretical possibility the reality often appears unchanged. For example, reviewing the literature on congenital heart disease Sullivan (2002) comments '... *survival advantage conferred by the prenatal diagnosis of congenital heart disease ... has been elusive*' although it is important to remember that conditions diagnosed prenatally may be more severe than those not identified in the prenatal period.

Decisions have to be made at different points through the process of prenatal testing, and at all of the decision points, as well as during the aftermaths, women and their partners will need specialised care and support to ensure they can understand the information they are given and make the right decisions for them. Provision of good care is both vital but difficult. Different aspects will be provided by different professionals, and that will vary across countries and how antenatal and pregnancy care is organised but some of the areas where it is particularly crucial include:

- ensuring informed choices are made at the time of undergoing screening
- ensuring parents understand the results of their screening test
- giving parents bad news
- supporting decision-making after a confirmed diagnosis of an abnormality
- supporting parents through termination of pregnancy
- supporting parents through a continued pregnancy
- provision of aftercare – post mortem/genetic information/support and/or counselling/next pregnancy.

To do all of these well, staff need to be trained and to feel supported.

6.3 Caring for Women and Observations Arising from the EDIG Study

Prenatal testing is a routine part of antenatal care for pregnant women across the EU. Different tests are offered in different countries and many different health professionals are involved in delivering these tests and in caring for women and their partners after positive or negative test results. Despite this lack of uniformity, we would like to suggest that there are standards of care that should be available for all pregnant women, and their partners. The findings from the EDIG study suggest that those standards are not always met. The important question of what tests are offered and related ethical issues were discussed in a previous paper (Statham and Dimavicius 2008). We would like to use the remainder of this chapter to raise some important issues that professionals can consider wherever they are practicing. We do not tell you what to do. What we hope to do is encourage you as health professionals to ask yourselves the question 'What is the optimum care that I can give?' with a particular focus on the ethical issues that you and the parents you care for may face during the process of prenatal screening, diagnosis and its aftermath.

6.3.1 Professional Values

Any professional working in antenatal screening and diagnosis may have a range of beliefs: that any choice a parent makes is acceptable; that screening contributes to reproductive choices; that it is acceptable for parents to end a pregnancy for a condition they feel is 'serious'. Others will be uncomfortable with aspects of screening and the decisions some parents make. In this area it is important that individuals' values do not influence the choices that parents make – as they will live with the consequences of whatever decisions they make. Professionals can inadvertently influence parents' decisions by their choice about how much information they give, how they describe a condition, the order in which they give information, the words they use, their body language and tone.

Each parent will bring their own different experiences, values and personal circumstances to the decisions they make during screening. To provide the best care for parents professionals must accept and respect their choices even if they do not sit easily with their own beliefs. Professionals who are aware of their own values can help parents explore their options with confidence across the whole process.

6.3.2 Informed Choice About Tests

The first decision for parents is whether or not to have the tests that are available to them in their system of care. For professionals this touches on two areas – their knowledge of tests and their skill in informing women and enabling them to make an informed choice.

6.3.2.1 The Offer of Tests – Professional Knowledge

Do parents know the differences between a screening test and a diagnostic test?
  Have you been trained to explain these differences to women?
  Are you confident that the screening tests you are offering are effective tests?

6.3.2.2 Telling Women About Screening Tests – Enabling Informed Choice

Is there written information available for women?
  Does the information you give tell women enough about the tests, the potential risks and benefits, the conditions being tested for, what happens next if this test shows there may be a problem and where to go for further information?
  Is there someone for women to talk to if they have further questions or concerns?
  Are women told about their choices early enough in pregnancy to be able to think about their decision to have screening and gather more information if they need it?

6.3.3 Giving Results

As well as talking to women and their partners about the tests, professionals have to give results. This may be the result of a blood test where a decision has to be made about further tests, or a finding from an ultrasound or other diagnostic test. News of even the possibility of something wrong with their baby will always be very hard for parents to receive.

6.3.3.1 Giving Screening Test Results – Planning for the Next Stage

Do you discuss with women how and when they will get their test results?

Do you make sure that women understand that a screen negative result does not guarantee a healthy baby?

Do you make sure that women understand that a screen positive result does not mean that it is certain that a baby has a problem?

Will there be someone available to talk to women who are anxious about their results?

How are parents helped to make an informed choice about any other tests they might be offered?

When women have ultrasound scans are they aware that it is both a screening and diagnostic test?

6.3.3.2 Giving a Diagnosis of a Fetal Abnormality – Compassionate Communication

This is the point at which parents' hopes and dreams of a healthy baby are taken away and replaced with news of a different baby.

If the woman has had an amniocentesis or chorionic villus sampling (CVS) do you discuss with her how and when she will get her test results?

When you tell parents this news do you have enough time

- to talk for as long as they need
- to see them again if they are too shocked to take in all of the news
- to explain fully what their options are and what they can expect?

Do you have enough information about the condition that has been diagnosed and do you know where else parents can go for more information or support, e.g. other professionals or groups supporting families with particular conditions?

When you have told the parents this news do you ensure that parents are clear about what will happen next?

6.3.4 Supporting Parents Making a Decision

This is a particularly difficult area for professionals and where the support you give has to be non-directive, non judgmental and free from your own values, attitudes and beliefs.

What is your role in helping parents as they make the decision whether to continue or terminate their pregnancy?

What information will parents need as they make their decision? What if the two partners have different views?

What other issues will parents be thinking about while they make their decision?

What do you do if parents ask 'what would you do?'

Parents need a framework in which to make decisions, as it may be the first time they have encountered such complex issues. Parents will need to explore their situation, the potential risks and benefits of alternative outcomes in order to reach a decision. You may have to raise some of the more challenging options with parents. As well as a structured approach focussing on information and process, parents will also use their prior beliefs and instincts as part of their decision-making.

For some parents, pragmatism, feelings and lived experience may lead them to make decisions which are not intellectually rigorous especially if the particular questions do not have definitive answers and this may be acceptable to them.

6.3.5 Helping Parents After They Have Made a Decision

Once an abnormality is confirmed, parents have lost the baby they had expected, the healthy baby. Whatever decision they make they have to live with the consequences in the short and long term. They need supportive care during either a termination or throughout the rest of the pregnancy.

6.3.5.1 Issues Around Termination

How do you help parents prepare for the practical and emotional aspects of ending a pregnancy and ensure that their care is sensitive and personal?

Are you aware of the paradox for parents who make the choice to have a termination but who are grieving for the loss of their child?

Can you help parents understand how the pregnancy will be terminated and where this will take place? Will there be pain relief? After an induced labour, do the parents know what to expect when the baby is born? Do you discuss whether they want to see and hold the baby, have photographs and memorials? Do you offer scan photographs to parents who terminate a pregnancy earlier under anaesthetic?

Do you discuss with parents how they might feel physically afterwards – pain, engorged breasts, bleeding? Do you discuss with parents how they might feel emotionally afterwards – and what might be available to help them?

6.3.5.2 Issues Around Continuing

What do parents need who continue their pregnancy after an abnormality has been diagnosed? The condition may be treatable or permanent and in many cases the parents may know that their baby will die before or soon after birth. Almost always the child who lives will have some physical and/or intellectual impairment. Parents will need on-going support as they may have many different worries and concerns as the pregnancy progresses such as

– Feelings about a 'different' pregnancy and a changed relationship with a different baby
– Fears and concerns about the practical aspects of their care
– Worries about the health of the baby
– The strain of adjusting to an ever changing scenario as the baby may need continuous monitoring throughout the pregnancy
– Preparing for the birth
– Preparing for afterwards: no baby OR the reality of an ill baby.

6.3.5.3 Longer Term Support for Parents and Families

Women who continue pregnancies may need to see a range of health, social and educational professionals depending on the condition of their child. They may also need psychological support around the emotional impact of a having a child with special needs.

Women whose babies die and those who have ended pregnancies may benefit from the offer of support from a practitioner who has been involved in their care and who can acknowledge their bereavement as well as help with any physical problems they may have. Some women will need more intensive psychological support and all professionals need to be sensitive to the indications for this especially when the need arises a long time after the termination.

6.3.6 Issues for Staff Themselves

To give the best care to parents, professionals need training. They will need to understand the whole process of prenatal testing in order to be sensitive to parents' experiences and needs. This will require specialised training for example, in the practical aspects of the tests, in communication, in breaking bad news and in bereavement support.

In an area that is technically complex and ethically challenging and that leads to witnessing the profound shock and grief that parents experience after a positive diagnosis, staff may need support. Often this is not available to staff but if they are to give high quality, sensitive and individualised care systems should be put in place for the support of staff. What provision is there in your institution to help you do this work effectively – to give parents the best care and to ensure your personal welfare?

6.4 Helping Professionals Deliver a Good Service

There are many Internet-based resources that health professionals will find helpful, both to use themselves and to recommend to parents for whom they are caring. There are many more sites but these are some of the English Language sites that the authors believe are useful and reputable. At the end of this section we also list a few papers that are particularly useful in considering ethical aspects of decision-making.

6.4.1 Sites and Organisations Which Give Information and Support to Parents

Antenatal Results and Choices http://www.arc-uk.org/

ARC is a UK-based charity which offers information and support to parents who are: making decisions before, during and after the antenatal testing process, told that their unborn baby has an abnormality, having to make difficult decisions about continuing the pregnancy or having to make difficult decisions about ending the pregnancy. The website describes the support available to parents, the training that is available for health professionals and also lists a range of organisations that support families affected by different conditions.

ANSWER aims to provide information to expectant parents and those thinking about becoming pregnant so that they can make the best decisions about prenatal testing, for themselves and their families, whether they decide on testing or not. The site includes personal testimonies from individuals and families directly affected by some of the conditions that can be tested for in pregnancy.

http://www.healthtalkonline.org/Pregnancy_children/Antenatal_Screening
http://www.healthtalkonline.org/Pregnancy_children/Ending_a_pregnancy_for_fetal_abnormality

Healthtalk Online shows a wide variety of personal experiences of health and illness and visitors to the site can watch, listen to or read interviews with people who have experienced a range of medical conditions or situations, as well as find reliable information on treatment choices and where to find support. The two modules identified above relate to prenatal testing and pregnancy termination.

6.4.2 Information from the UK National Screening Committee

UK National Screening Committee

The link http://www.screening.nhs.uk/cpd is the UK National Screening Committee's Continuing Professional Development website for England. It is a 'one stop' website bringing together all the cross-cutting education and training programmes and resources – meeting the continuing professional development (CPD) needs of health professionals involved in NHS antenatal and newborn screening.

Of particular interest is Screening Choices: http://cpd.screening.nhs.uk/screeningchoices

This learning programme aims to enhance skills and knowledge of health professionals in order to ensure women and families can make informed choices about offers of screening. Online materials enable professionals to work through modules on all aspects of screening during pregnancy and in the neonatal period.

A range of information for women can be found at http://fetalanomaly.screening.nhs.uk/publicationsandleaflets and of particular interest is the link to one called '*Screening tests for you and your baby*'. The first page of this 72 page booklet makes it clear what it aims to do:

> This booklet is about the screening tests you will be offered in pregnancy and screening for your baby in the first few weeks after birth. It is important you understand the purpose and possible results of the screening tests before you make your decision. To help you the UK National Screening Committee (the independent body that advises the health departments of the four UK countries) has written this booklet, explaining the screening tests in detail.
>
> We realize you are given a lot of information in pregnancy but please read this booklet as it will help prepare you for discussions with your midwife or doctor about the screening tests, so you can ask the questions that are important to you. It will be helpful if you have the booklet with you when you see them. As you can see from the screening timeline diagram on the inside front cover, some of the tests need to take place as early as 10 weeks in pregnancy so we recommend reading the booklet as soon as possible. Towards the end of your pregnancy your midwife will discuss with you the screening tests recommended for newborn babies. We advise you to look at the booklet again at this stage.

6.4.3 Information About Organisations Dealing with a Broad Range of Genetic Conditions

The Genetic Interest Group http://www.gig.org.uk/

The Genetic Interest Group (GIG) is a national alliance of patient organisations with a membership of over 130 charities which support children, families and individuals affected by genetic disorders. GIG is a organisation which aims to improve the lives of individuals and families affected by genetic disorders through working within the UK and European Union at policy and legislative level.

Genetic Alliance UK is the national charity of patient organisations with a membership of over 130 charities supporting all those affected by genetic disorders. The aim of the organisation is to improve the lives of people affected by genetic conditions by ensuring that high quality services and information are available to all who need them

CAF is the UK-wide charity providing advice, information and support to the parents of all disabled children. CAF maintains an A–Z of more than 1,200 conditions with medical information on disabilities and rare disorders and details of support groups.
http://www.cafamily.org.uk/medicalinformation/conditions/azlistings/a.html
Contact a Family's vision is that all families with disabled children are empowered to live the lives they choose to live, and achieve their full potential, for themselves, for the communities they live in and for society.
Contact a Family's mission and purpose is to remove the barriers imposed by society which prevent families with disabled children achieving their full potential, and to empower these families to live the lives they want to lead.

6.4.4 European Organisations

EURORDIS

EURORDIS http://www.eurordis.org/secteur.php3?id_rubrique=1

EURORDIS is a non-governmental patient-driven alliance of patient organisations and individuals active in the field of rare diseases, dedicated to improving the quality of life of all people living with rare diseases in Europe. It was founded in 1997; it is supported by its members and by the French Muscular Dystrophy Association (AFM), the European Commission, corporate foundations and the health industry.

A selection of the information is translated into French, German, Italian, Portuguese and Spanish.

http://www.eurogenguide.org.uk/

EuroGenGuide contains information about genetic testing, counselling and research across Europe. The information is intended as a resource for both patients and their families, and health professionals alike. You might be concerned that you or a relative may be affected by a genetic condition, or your doctor may have asked you if you would be prepared to donate biological samples to a DNA bank for use in research. Alternatively you may be a doctor with a patient who might have a genetic condition and requires a test, or you might be a genetic researcher carrying out a study.

6.4.5 Useful References Addressing Ethical Issues Around Decision-Making in a Practical Way

ARC (2005) Supporting parents' decisions: a handbook for professionals. Antenatal Results and Choices, London

Bayliss F, Downie J (2001) Professional recommendations: disclosing facts and values. J Med Ethics 27(1):20–24

Berkel van D, Weele van der C (1999) Norms and pre-norms on prenatal diagnosis: new ways to deal with morality in counselling. Patient Educ Couns 37:153–156

O'Connor AM, Bennett CL, Stacey D, Barry M, Col NF, Eden KB, Entwistle VA, Fiset V, Holmes-Rovner M, Khangura S, Llewellyn-Thomas H, Rovner D. Decision aids for people facing health treatment or screening decisions. *Cochrane Database of Systematic Reviews* 2009, Issue 3. Art. No.: CD001431. DOI: 10.1002/14651858.CD001431.pub2

6.5 Conclusion

Prenatal testing is widely available in Europe and the developed world. How tests are delivered to pregnant women is dependent on how maternity care is organised and financed, on how health care is regulated and also on existing legal frameworks. We have discussed these issues in a previous paper. The emphasis in this chapter is on an issue of overriding concern: the way in which women are cared for during the process of prenatal testing. Testing involves women and their partners making difficult decisions about issues which are often ethically challenging and may carry some risks to their unborn baby. Appropriate care recognises the need for health professionals, as well as women and their partners, to be informed, to respect different perspectives and to understand the situation in which parents find themselves. Appropriate care can however only be given by well trained and well supported health professionals.

References

Bourne P (1968) The psychological effects of stillbirths on women and their doctors. J R Coll Gen Pract 16:103–112

Bui T-H, Meiner V (2008) State of the art in prenatal diagnosis. In: Leuzinger-Bohleber M, Engels E-M, Tsiantis J (eds.) The janus face of prenatal diagnostics. Karnac, London, pp 61–86

Carrapato MR, Marcelino F (2001) The infant of the diabetic mother: the critical developmental windows. Early Pregnancy 5(1):57–58

Evans MI, Pryde PG et al (1993) The choices women make about prenatal diagnosis. Fetal Diagn Ther 8:70–80

Fairgrieve SD, Jackson M et al (2000) Population based, prospective study of the care of women with epilepsy in pregnancy. Br Med J 321:674–675

Fischmann T, Pfenning N et al (2008) Empirical data evaluation on EDIG. In: Leuzinger-Bohleber M, Engels E-M, Tsiantis J (eds.) The janus face of prenatal diagnostics. Karnac, London, pp 89–136

Fuchs F, Riis P (1956) Antenatal sex determination. Nature 177:330

Holmes LB, Harvey EA et al (2001) The teratogenicity of anticonvulsant drugs. N Engl J Med 344(15):1132–1138

Kennell JH, Slyter H et al (1970) The mourning response of parents to the death of a newborn infant. N Engl J Med 283:344–349

Kessler S (1992) Psychological aspects of genetic counseling. VII. Thoughts on directiveness. J Genet Couns 1:164–171

LeJeune J, Gautheir M et al (1959) Les chromosomes humains en culture de tissus. Comptes Rendu Acad Sci 248:602–603

Lowe SA (2001) Drugs in pregnancy. Anticonvulsants and drugs for neurological disease. Baillières Best Pract Res Clin Obstet Gynaecol 15(6):863–876

Makowski EL, Prem KA et al (1956) Detection of sex of fetuses by the incidence of sex chromatin body in nuclei of cells in amniotic fluid. Science 123:542–543

Moore KL, Graham M et al (1953) The detection of chromosomal sex in hermaphrodites from a skin biopsy. Surg Gynecol Obstet 96:641–648

Nadler HL (1968) Antenatal detection of hereditary disorders. Pediatrics 42:912–918

Parkes CM (1965) Bereavement and mental illness. Part 1. A clinical study of the grief of bereaved psychiatric patients. Br J Med Psychol 38:1–12

Parkes CM (1995) Guidelines for conducting ethical bereavement research. Death Stud 19:171–181

RCP (1989) Prenatal diagnosis and screening, A report of the Royal College of Physicians

Resta RG (1997) The first prenatal diagnosis of a fetal abnormality. J Genet Couns 6(1):81–84

Riis P, Fuchs F (1960) Antenatal determination of foetal sex in prevention of hereditary diseases. Lancet ii:180–182

Serr DM, Sachs L et al (1955) The diagnosis of sex before birth using cells from the amniotic fluid. Bull Res Counc Isr 5B:137–138

Shettle LB (1956) Nuclear morphology of cells in human amniotic fluid in relation to sex of infant. Am J Obstet Gynecol 71:834

Smoak IW (2002) Hypoglycemia and embryonic heart development. Front Biosci 7:d307–d318

Statham H (1992) Professional understanding and parents' experience of termination. In: Brock DJH, Rodeck CH, Ferguson-Smith MA (eds) Prenatal diagnosis and screening. Churchill Livingstone, London, pp 697–702

Statham H (1994) Parents' reactions to termination of pregnancy for fetal abnormality: from a mother's point of view. In: Abramsky L, Chapple J (eds) Prenatal diagnosis: the human side. Chapman & Hall, London, pp 157–172

Statham H (2002) Prenatal diagnosis of fetal abnormality: the decision to terminate the pregnancy and the psychological consequences. Fetal Matern Med Rev 13:213–247

Statham H, Dimavicius J (2008) Ethics of care in prenatal diagnosis: implications of variations in law, policy and practice in EDIG countries. In: Leuzinger-Bohleber M, Engels E-M, Tsiantis J (eds.) The janus face of prenatal diagnostics. Karnac, London, pp 45–60

Statham H, Solomou W et al (2000) Prenatal diagnosis of fetal abnormality: psychological effects on women in low-risk pregnancies. Baillières Best Pract Res Clin Obstet Gynaecol 14(4):731–747

Steele MW, Breg WR (1966) Chromosome analysis of human amniotic fluid cells. Lancet 1:383–385

Sullivan ID (2002) Prenatal diagnosis of structural heart disease: does it make a difference to survival? Heart 87(5):405–406

Wertz DC, Fletcher JC (1989) Ethical issues in prenatal diagnosis. Pediatr Ann 18:739–749

Wong SF, Chan FY et al (2002) Routine ultrasound screening in diabetic pregnancies. Ultrasound Obstet Gynecol 19(2):171–176

Chapter 7
Cooperation Is Rewarding If the Boundary Conditions Fit: Interdisciplinary Cooperation in the Context of Prenatal Diagnostics

Astrid Riehl-Emde, Anette Bruder, Claudia Pauli-Magnus, and Vanessa Sieler

Abstract In Germany, pregnant women have a legal right to psychosocial care in the context of prenatal diagnostics (PND); however, this is not widely known and only a few women take advantage of it. In an attempt to improve the situation, various model projects have been implemented in recent years to promote cooperation between medical and psychosocial professional groups. The Institute of Psychosomatic Cooperation Research and Family Therapy at the University Hospital Heidelberg was previously responsible for conducting the accompanying research in two model projects and is currently conducting a third. This paper summarizes results from approximately 10 years of continuous research in this area. The focus here is on the aspects of collaboration between the professional groups involved within the context of PND. On the whole, there have been definite advances with regard to the criteria for care, interdisciplinary cooperation, continuing education and advanced training in the context of PND as well as school projects and outreach work, but development in this field is moving rather slowly. We concur that, as previously recommended by professionals and politicians, it is time to coordinate information and care for pregnant women and their partners.

Keywords Prenatal counselling • Prenatal diagnostics • Psychosocial aspects

7.1 Introduction

In Germany, since 1976, prenatal diagnostics (PND) is included in the catalogue of services paid by medical insurance funds. Since then, the technical possibilities in the field of PND have developed further. Today, PND has become a routine part of

A. Riehl-Emde (✉), A. Bruder, C. Pauli-Magnus, and V. Sieler
Institute of Psychosomatic Cooperation Research and Family Therapy,
Centre for Psychosocial Medicine, University Hospital Heidelberg, Bergheimer Straße 54,
D-69115 Heidelberg, Germany
e-mail: astrid.riehl-emde@med.uni-heidelberg.de

medical practice during prenatal care. Thus, prenatal care, in addition to monitoring the general course of pregnancy, includes tests which can detect fetal abnormalities.

Prenatal diagnostics today include the following prenatal tests:

- General prenatal care (ultrasound screening);
- Methods of estimating risk/screening tests (nuchal translucency screening, first trimester test, triple test);
- Invasive/diagnostic methods (chorionic villus sampling, testing of amniotic fluid, fetal blood sampling from the umbilical cord).

German medical insurance funds currently pay only for services that are medically necessary and appropriate. In addition to general prenatal care, this includes the follow-up examination when an abnormal finding has been obtained. Methods of risk appraisal are offered as individual health services ("Individuelle Gesundheitsleistungen", IGeL) and are therefore billed to the patients themselves. This does not apply, if the test shows an abnormal result in which case the investigation is considered 'medically necessary' and, thus, paid for by the insurance.

The diagnostic possibilities currently exceed the therapeutic possibilities because intrauterine treatment is only possible for very few diagnoses. The indisputable advances in new techniques certainly can offer a great deal of reassurance for the mother about her unborn child during pregnancy. However, these advantages must be seen against the psychological stresses experienced by women and couples when they are informed about an unclear or abnormal finding. Ultimately, the parents-to-be are confronted with the ethically challenging decision to terminate the pregnancy or to intentionally decide to give birth to, and perhaps live with, a disabled child. This has been referred to as an "impossible decision" (Friedrich et al. 1998). Problems that were previously left to fate have thus come within the individual decision-making realm and make high demands on individual decision-making competency. Many people are stressed by this, not only because their competencies are not sufficient but also because this involves serious ethical dilemmas and may involve questions that go beyond human decision-making capacity in general.

To deal with these problems, the legal right to counselling has been anchored in the context of PND with §2 of the German Pregnancy Conflict Act ("Schwangerschaftskonfliktgesetz", SchKG). The uptake of counselling has so far been voluntary. The counselling should be nondirective with regard to outcome and should address the ethical and psychosocial dimensions of PND. Embedding PND into a legal act has contributed significantly towards the establishment of PND since the 1970s. To be able to offer a corresponding counselling infrastructure, human genetic counselling offices were expanded and additional services were created within existing psychosocial counselling offices, administered publicly or by communities. However, only a few women take advantage of the legal right to counselling and most women are not even aware of it. Furthermore, cooperation between physicians and psychosocial professionals rarely takes place and has proven to be extremely difficult, especially outside of university facilities. Therefore, various model projects were initiated. Two of these projects with which the above mentioned Heidelberg Institute was involved are reported in this article. In addition, a summary of additional research in this field covering 10 years is presented. The focus here is on the features of cooperation between the professional groups involved in the context of PND.

7.2 Accompanying Research at the Institute of Psychosomatic Cooperation Research and Family Therapy, University Hospital Heidelberg

The Institute of Psychosomatic Cooperation Research and Family Therapy has more than 10 years' expertise with respect to conducting model projects within the context of PND. The first was the project "Development of Counselling Criteria for Prenatal Counselling in Cases of Anticipated Disability of the Child" (1998–2001), which was supported by the Federal Ministry of Family Affairs, Senior Citizens, Women and Youth ("Bundesministerium für Familie, Senioren, Frauen und Jugend", BMFSFJ 2001). This was followed by "Interprofessional Quality Circle in Prenatal Diagnostics," supported by the Federal Centre for Health Education (2003–2008) ("Bundeszentrale für gesundheitliche Aufklärung", BZgA 2008). In addition, since 2008, the Heidelberg Institute has been involved in research supported by the Federal German State of Baden-Württemberg with the aim of (i) improving prenatal information about PND and (ii) improving counselling and support of parents-to-be in conflict situations during and after prenatal diagnostic tests (Pauli-Magnus et al. 2011).

7.3 Model Project "Development of Counselling Criteria for Prenatal Counselling in Cases of Anticipated Disability of the Child"

The initial aim of the project "Development of Counselling Criteria for Prenatal Counselling in Cases of Anticipated Disability of the Child" was to develop counselling criteria and thus improve the quality of outcome and the process of psychosocial counselling in PND. However, the analysis of shortcomings in psychosocial care for pregnant women (Table 7.1) led to changes and additions in the first year of the project.

This analysis revealed a lack of information for pregnant women about what was available to them, and insufficient cooperation between the psychosocial counselling offices and medical institutions, in particular gynaecological practices. Such interdisciplinary cooperation was explicitly requested in the guidelines of the German Federal Chamber of Physicians (Bundesärztekammer 1998), but, in practice, this rarely took place. Therefore, the following three main topics of activity were created: "counselling criteria", "advanced training and ongoing education" and "cooperation" (Lammert et al. 2002).

Table 7.1 Deficits in structure quality from the perspective of patients/clients, gynaecologists and counselling offices

Patients/clients
Too little acceptance and utilization of the offer of counselling due to:
　Lack of awareness of the offer
　Uncertainty due to inconsistent information
　High inhibition threshold against receiving psychosocial counselling
Gynaecologists
Too little information about the counselling offerings due to:
　Different understanding of counselling
　Inadequate or insufficient information about the psychosocial counselling offering
　Doubt about the confidence of the other professional group
　Feeling of responsibility for one's own patients
Counselling offices
　Unclear counselling profile
　Fears and anxieties with regard to the dominance of medicine

7.4 Model Project "Interprofessional Quality Circle in Prenatal Diagnostics"

The model project "Interprofessional Quality Circle (IQC) in Prenatal Diagnostics" was conducted in order to improve the quality of counselling in the context of PND. The aim of the project was to promote cooperation between medical and psychosocial professional groups in accordance with the concept of Bahrs et al. (2001, 2005). To do so, IQCs were initiated with a time lag over two phases at six sites throughout Germany (at Augsburg, Erfurt, Freiburg, Heidelberg, Mannheim, and Schwerin), where gynaecologists, human geneticists, paediatricians as well as consultants from pregnancy counselling offices and midwives were able to participate. The circles were each moderated by a specifically trained person from both the medical and psychosocial professional group. According to Bahrs's concept, the participants would present typical work situations and problems based on case presentations from their own work, which would then be analyzed in the group. On this basis, the group should work continuously on proposed improvements for the counselling and should adhere to them in recommendations for action developed together.

The results show that medical and psychosocial group members felt better informed about what the other group could offer during the IQC work. An improved perception of each other was reported, in particular a better mutual understanding, more openness and transparency as well as a better interdisciplinary cooperation. At five of the six sites, the IQC work continued beyond the model project phase; four circles were still active at the time of the follow-up, i.e., 2–4 years after the end of the project. It was also reported that physicians referred their patients more often to psychosocial counselling. Furthermore, outpatient psychosocial offices were set up at hospitals at two sites. At all sites, there were more public awareness activities.

Surprisingly, however, there was no increase in PND counselling at the counselling services (Riehl-Emde et al. 2007; Kuhn and Riehl-Emde 2007; Kuhn et al. 2008).

7.5 Cooperation Is Worthwhile If the Boundary Conditions Fit

As part of the follow-up, the moderators from the two professional groups were asked in semi-standardized interviews about their subjective evaluation of the IQC work and its effects on interdisciplinary cooperation. Interviews were transcribed and a content analysis according to Mayring (2003) was undertaken.

With regard to changes due to the quality circle work, there is a spectrum of views. Some sites had established various methods of cooperation and were still highly motivated to cooperate even after the end of the model project. Other sites had achieved few changes with regard to interdisciplinary cooperation and did not continue the IQC work after the project has ended.

In a subsequent step, and based on the above two groups, interview material was analysed for specific relationships at a particular site. The results show that achieved improvements mainly depend on the interpersonal experience and the individual benefit from the IQC work. Every site can be assigned to one of two groups (Table 7.2); the left column in the table lists the sites where positive interpersonal experience and moderate to great changes have been described; and the right column lists sites where negative interpersonal experience and few changes or none at all were described. The assignment was based on a detailed analysis of particular "linguistic markers of meaning" which provide information about how intensely developments took place. The positive and negative interpersonal experience was associated with a high and a low personal benefit, respectively. This factor had a significant influence on the extent of the individual involvement.

If the external and personal boundary conditions fit, interdisciplinary cooperation has the following advantages for the medical community:

– Increased attractiveness of a practice/department because of greater patient satisfaction (Rohde and Woopen 2007);
– Emotional relief due to shared responsibility for patients.

Table 7.2 Interpersonal experience, personal benefit and work on contents as a function of the extent of change

Positive interpersonal experience	Negative interpersonal experience
Assessment of value	Inadequate assessment of value
Confidence	Competition, rejection
Openness	Inadequate openness
Engagement	Inadequate engagement
Motivation	Inadequate motivation
Competence	Inadequate competence
High personal benefit	Lower personal benefit
Intense work on content	Little work on content

Interdisciplinary cooperation offers the following advantages for the professional psychosocial group in particular:

- An opportunity for networking and better positioning in the field of PND;
- Increased demand for the offered psychosocial information and counselling services in the context of PND.

7.6 Retrospective

For more than 10 years, work has been conducted into improving overall counselling and cooperation within the context of PND, not just at the Heidelberg institute but also at other research centres. In comparison with the situation at the end of the 1990s, essential progress has been made in the following three fields of work:

- *Counselling criteria*: Initially there were only guidelines for medical counselling in the context of PND (Bundesärztekammer 1998) and also recommendations for psychosocial counselling (Lammert and Neumann 2002). Additional recommendations on how to encourage cooperation between medical and psychosocial counselling have now been made available and tested as part of the IQC project (BZgA 2008).
- *Advanced training and continuing education*: A curriculum for the joint advanced training and continuing education of the medical community and psychosocial consultants for counselling in conjunction with PND was developed, tested and evaluated[1] and has been established at the Central Evangelical Institute for Family Counselling (ezi) in Berlin.
- *Cooperation*: The following models for interdisciplinary cooperation are currently practiced, depending on local conditions: (i) traditional referral, i.e. a definitive recommendation to a facility and/or psychosocial consultant; (ii) office hours "on call" at the medical facility, i.e. the counselling personnel only becomes active on request; (iii) medical and psychosocial counselling "under one roof". Interprofessional quality circles represent an additional model of cooperation. This model contrasts with the forms of cooperation mentioned above, because the participating professional groups exchange information about their work with each other and not in direct contact with patients. – From the standpoint of the physicians, "under one roof" model is probably the most promising. This was shown in the model project "Psychosocial Counselling in the Context of PND" (Rohde and Woopen 2007) that was conducted in the federal German state of North Rhine-Westphalia. It showed that pregnant

---

[1]Evangelisches Zentralinstitut für Familienberatung gGmbH (Ed.) *Abschlussbericht zum Modellprojekt "Entwicklung, Erprobung und Evaluation eines Curriculums für die Beratung in Zusammenhang mit vorgeburtlichen Untersuchungen (Pränataldiagnostik) und bei zu erwartender Behinderung des Kindes (2002 bis 2005)"*, supported by BMFSFJ; www.bmfsfj.de/Kategorien/Forschungsnetz/Forschungsberichte.html

women with abnormal PND findings, who underwent psychosocial counselling in immediate proximity to medical care, had a higher acceptance and utilization of it. Apparently, such immediate proximity facilitates short-term agreements between the professional personnel and creates fewer barriers to uptake by the patients. To be perceived as a second opinion by the patient, psychosocial counselling has to present an independent view and should, ideally, belong to another institution.

In addition, there are other areas of work such as:

- *School projects*: In addition to a number of other projects ("Overview of ongoing and recently concluded research projects on the theme complex of prenatal diagnostics": http://pnd-tagung.sexualaufklaerung.de/index.php), it should also be pointed out that efforts have already begun in some cities to discuss questions associated with the importance of pregnancy and child rearing in the schools. For example, in 2006, the Interdisciplinary Forum for Biomedicine and Cultural Sciences (IFBK) at the University of Heidelberg began a school project[2] concerning stem cells and clones, prenatal diagnostics and abortion. This project "Human Dignity at the Beginning of Life" was conducted at four high schools in Heidelberg, Mannheim and Karlsruhe in the school year 2007/2008.
- *Public Relations*: The BZgA has established essential cornerstones through public relation, information materials (brief flyers with initial information, brochures with more in-depth content and electronic Information http://www.schwanger-info.de) and, not least of all, through representative interviews of pregnant women about "Pregnancy experiences and prenatal diagnostics" (BZgA 2006).

7.7 Current Situation

Although there have been various activities on different levels extending all the way to model projects as well as an increase in the willingness of various professional groups to cooperate, development in the field has been relatively slow. Progress in interdisciplinary cooperation in the context of PND has been relatively minimal, at least when measured solely by the number of cases, i.e., the number of pregnant women receiving psychosocial counselling. In the last 10 years, work has been done with the particular intention of improving the cooperation structures, the further training of technical personnel and the development of counselling criteria, with some success. This success could, however, have been greater in view of the time, personnel and financial resources invested. On the whole, the professional meeting about "Cooperation – But How? Interdisciplinary Cooperation in Prenatal Diagnostics" which was initiated by the BZgA in Berlin in May 2007, has shown

---

[2] Supported by the Robert Bosch Foundation.

that discussion about cooperation structures is far from over and struggles to improve the quality of the structures still take place.

In the meantime, not only political groups and associations but also, to an increasing extent, representatives of the medical community have been concerned that pregnant women rarely avail themselves of their legal right to counselling within the context of PND and are often completely unaware of this legal right. The level of information pregnant women have regarding PND is relatively low (BZgA 2006).

As a result of our work we would propose that the pregnant women, in whose interest such model projects have been conducted, must become more involved. The fact that pregnant women often decline psychosocial counselling, despite medical recommendation, has previously been attributed, mainly, to the spatial separation between medical practice and psychosocial counselling facilities. In fact, more pregnant women undergo counselling if the two facilities are "under one roof." However, it is conceivable that pregnant women are rejecting psychosocial counselling for other reasons. It is, for example, known that they tend to avoid difficult and/or ambivalent topics. Such behaviour may be understood as self-protection in the interest of maintaining an optimistic, relaxed basic attitude (Brisch 2007). Why should a pregnant woman open herself up to the conflicts associated with PND, if these might not be relevant to her and so deny her a positive experience of pregnancy? Studies of the decision-making processes in pregnant women have also shown that information about the risks of PND is usually accepted better than information about the conflict potential (Nippert and Neitzel 2007). Psychoanalytical experience indicates that PND may serve as a projection screen for aggressive and destructive fantasies. Such projections may help to experience the pregnancy with as little ambivalence as possible, given that a pregnancy is always associated with uncertainties and anxieties (Belz 2008). We would recommend there should be more consideration of the psychosocial situation of the pregnant women. Efforts should be made to discover how it may be possible to support a relaxed and optimistic attitude in pregnant women and to reach them in their specific emotional world but at the same time to provide them with sufficient information without creating resistance, so they can benefit from the technical options of PND.

7.8 Prospects

In the year 2008, on the initiative of the Labour and Social Ministry of the Federal German State of Baden-Württemberg, and with the support of representatives of the medical community in that state, another model project was created, and again the above mentioned Heidelberg Institute was selected to conduct the research.[3]

---

[3] Project director (Prof. Astrid Riehl-Emde, Ph.D.) in cooperation with the Department for General Gynecology and Obstetrics (medical director: Prof. Christof Sohn, M.D.), likewise at the Heidelberg University Clinic.

The first aim of this project is to improve early information about PND for pregnant women, namely *before* they avail themselves of prenatal diagnostic measures, and secondly, to improve the counselling and support of parents-to-be in conflict situations *during* and *after* PND. As part of this project, first the existing structures and measures at eight sites in Baden-Württemberg (at Balingen, Böblingen, Heilbronn, Karlsruhe, Constance, Mannheim, Stuttgart, and Ulm) were expanded and intensified. In the next stage, the focus will be on the situation of pregnant women. Two interviews with pregnant women about their "experiences of pregnancy and prenatal diagnostics" are scheduled: first, pregnant women who have received counselling at the counselling offices within the context of PND are to be interviewed, and secondly, the representative interview of pregnant women is to be repeated, including in-depth interviews in a subgroup of pregnant women to ascertain whether their information level is higher in the year 2010 than it was in the year 2004 (BZgA 2006).

Scientific progress in the various fields of medicine – theoretical as well as diagnostic and therapeutic – has touched on the question of whether everything which is medically possible should also be implemented. The question is, whether medical intervention, attempted and performed in the interest of reducing suffering, will bring more benefit than harm in the medium- and long-term-run. Can experts define the indications for termination of pregnancy and can medical doctors implement this without crossing personal lines? Even if they cannot provide answers to all these questions, model projects may offer an opportunity for various professional groups active in the context of PND to jointly search for ways to deal with the new findings and the ethical challenges associated with PND.

References

Bahrs O, Gerlach FM, Szecsenyi J, Andres E (eds.) (2001) Ärztliche Qualitätszirkel. Leitfaden für den Arzt in der Praxis und Klinik, 4th edn. Deutscher Ärzteverlag, Köln

Bahrs O, Nave M, Zastrau B (2005) Was sind Qualitätszirkel? Konzeptuelle Grundlagen und Abgrenzung von anderen Formen der Gruppenarbeit. In: Bundeszentrale für gesundheitliche Aufklärung (BZgA) (ed.) Qualitätszirkel in der Gesundheitsförderung und Prävention – Handbuch für Moderatorinnen und Moderatoren. BZgA, Köln, pp 23–53

Belz A (2008) "I'd also like to be in good hope myself for once." The highly problematic decision-making process within the framework of PND and its dependency on a sufficiently developed, autonomous female identity. In: Leuzinger-Bohleber M, Engels E-M, Tsiantis J (eds.) The janus face of prenatal diagnostics: a European study bridging ethics, psychoanalysis, and medicine. Karnac, London

Brisch KH (2007) Angst und Bewältigungsformen von Schwangeren und kindliche Entwicklung bei pränataler Ultraschall-Diagnostik. Prax Kinderpsychol Kinderpsychiatr 56:795–808

Bundesärztekammer (1998) Richtlinien zur pränatalen Diagnostik von Krankheiten und Krankheitsdispositionen. Dtsch Ärztebl 95(50):A3236–A3242 (most recent release 2003)

Bundeszentrale für gesundheitliche Aufklärung (BZgA) (2006) Schwangerschaftserleben und Pränataldiagnostik. Repräsentative Befragung Schwangerer zum Thema Pränataldiagnostik. BZgA, Köln

Bundeszentrale für gesundheitliche Aufklärung (BZgA) (2008) Interprofessionelle Qualitätszirkel in der Pränataldiagnostik. Forschung und Praxis der Sexualaufklärung und Familienplanung, vol 30. BZgA, Köln

Bundesministerium für Familie, Senioren, Frauen und Jugend (BMFSFJ) (Ed.) (2001) Materialien zur Familienpolitik. Abschlussbericht des Modellprojekts "Entwicklung von Beratungskriterien für die Beratung Schwangerer bei zu erwartender Behinderung des Kindes". Bonn

Friedrich H, Henze K-H, Stemann-Acheampong S (1998) Eine unmögliche Entscheidung. Pränataldiagnostik: Ihre psychosozialen Voraussetzungen und Folgen. VWB, Berlin

Kuhn R, Riehl-Emde A (2007) Gute Voraussetzung für eine Zusammenarbeit zwischen Ärzten und psychosozialen Fachkräften in der Beratung zur Pränataldiagnostik. Psychother Psychosom Med Psychol 57:44–52

Kuhn R, Bruder A, Cierpka M, Riehl-Emde A (2008) Beratung zu Pränataldiagnostik – Einstellungen und Haltungen von ärztlichen und psychosozialen Fachkräften. Ergebnisse aus dem Modellprojekt "Interprofessionelle Qualitätszirkel in der Pränataldiagnostik". Geburtshilfe Frauenheilkd 68:172–177

Lammert C, Neumann A (2002) Beratungskriterien. In: Lammert C, Cramer E, Pingen-Rainer G, Schulz J, Neumann A, Beckers U, Siebert S, Dewald A, Cierpka M (eds.) Psychosoziale Beratung in der Pränataldiagnostik – Ein Praxishandbuch. Hogrefe, Göttingen, pp 45–96

Lammert C, Cramer E, Pingen-Rainer G, Schulz J, Neumann A, Beckers U, Siebert S, Dewald A, Cierpka M (eds.) (2002) Psychosoziale Beratung in der Pränataldiagnostik – Ein Praxishandbuch. Hogrefe, Göttingen

Mayring P (2003) Qualitative Inhaltsanalyse. Grundlagen und Techniken, 8th edn. Beltz, Weinheim

Nippert I, Neitzel H (2007) Ethische und soziale Aspekte der Pränataldiagnostik. Prax Kinderpsychol Kinderpsychiatr 56:758–771

Pauli-Magnus C, Bruder A, Sieler V, Engelken-Juki S, Riehl-Emde A (2011) Interprofessionelle Vernetzung im Kontext von Pränataldiagnostik. Eine qualitative Studie über Erfahrungen von Beraterinnen im Rahmen eines Modellprojekts. Familiendynamik 36(1):32–42

Riehl-Emde A, Kuhn R, Dewald A, Cierpka M (2007) Interprofessionelle Qualitätszirkel in der Pränataldiagnostik: Ein Modellprojekt verbessert die Versorgung. BZgA Forum Sexualaufklärung Familienplanung 1:21–25

Rohde A, Woopen C (2007) Psychosoziale Beratung im Kontext von Pränataldiagnostik. Evaluation der Modellprojekte in Bonn, Düsseldorf und Essen. Deutscher Ärzteverlag, Köln

Chapter 8
Prenatal Genetic Counselling: Reflections on Drawing Policy Conclusions from Empirical Findings

Anders Nordgren

Abstract In this chapter I discuss the implications of two major empirical findings of the EU-wide project "Ethical Dilemmas due to Prenatal and Genetic Diagnostics" (EDIG) for a professional policy of prenatal genetic counselling. I argue that these empirical findings do not in and by themselves say anything about what such a policy should look like. Their significance depends on ethical and other presumptions already accepted within a particular policy framework. However, the empirical findings may also be significant in another more indirect way. They may show the inadequacy of assumptions about facts about the attitudes and views of women undergoing prenatal testing, and thereby undermine the policy that these assumptions about facts are used to support. In this way the findings may give proponents of the policy reason to reconsider their view.

Keywords Ethics • Genetic counselling • Prenatal testing • Policy frameworks

What are the implications of the empirical findings of the EDIG project for an ethically acceptable professional policy of prenatal genetic counselling? Most people would probably agree that the findings provide a better understanding of the psycho-social situation, attitudes and views of women undergoing prenatal screening and diagnosis, and that this understanding may inform our reflection on policy issues around prenatal genetic counselling. However, what does it mean more precisely to take these findings seriously? Do they say anything about what a policy of prenatal genetic counselling *should* look like? In this paper, I discuss problems that arise when we try to draw normative conclusions from these empirical findings.

I focus on four questions:

- What are the most important empirical findings of the EDIG project possibly relevant to prenatal genetic counselling?
- What are the main optional policy frameworks for prenatal genetic counselling?

A. Nordgren (✉)
Centre for Applied Ethics, Linköping University, SE-58183 Linköping, Sweden
e-mail: anders.nordgren@liu.se

- What do these normative and evaluative frameworks imply regarding the significance of the empirical findings?
- How may the empirical findings undermine a particular policy framework?

8.1 Empirical Findings

Let me start by very briefly presenting two major empirical findings of the EDIG project that could be relevant to prenatal genetic counselling.

The first finding concerns how women view the role of health care professionals. Sixty-eight percent of the women who received a test result indicating a problematic finding were very satisfied or satisfied that the health professionals left the decisions to them. The decision could be whether or not to have the test in the first place, or whether or not to terminate the pregnancy due to the test results. Some women wanted some limited advice. A minority wanted more advice (Fischmann et al. 2008, p. 121).

The second major finding is that decision-making in this context is not merely an intellectual process, but a highly emotional one. Distress is high when women make a decision to terminate their pregnancy, but it diminishes over time for most women. Some women, however, remain distressed for some time afterwards (Fischmann et al. 2008, pp. 108–117).

It should be noted that the statistical basis of these findings is rather limited. They concern only N women in N countries (Fischmann et al. 2008).

8.2 Normative Policy Frameworks for Prenatal Genetic Counselling

In analysing optional policies of prenatal genetic counselling, I find it useful to make a distinction between policy frameworks and more specific goals (for a more extensive analysis, see Nordgren 2008). As I use the term, a "policy framework" is a framework for health professionals involved in genetic counselling. It focuses on core values that determine which more specific goals are to be attained. Policy frameworks are normative as well as evaluative. They are normative, because they prescribe how genetic counselling should be carried out. They are evaluative, because they are centred around a particular key value that is to be realised. Policy frameworks may include ethical presumptions, but also, e.g., presumptions regarding women's psychological or social needs.

I distinguish at least three types of policy frameworks (Nordgren 2008):

- client-centred frameworks,
- health-centred frameworks, and
- anti-abortion frameworks.

These frameworks differ regarding whose perspective should be taken into account by the health professionals. Should the focus be on the woman or on the fetus?

Client-centred frameworks focus on the woman and aim at satisfying her preferences and/or needs. Given this central value, clients should not be recommended what to decide, i.e., whether or not to have a test or whether or not to terminate the pregnancy on the basis of test results. Non-directiveness regarding the content of a decision is essential. According to Wertz, this view dominates among health professionals in English-speaking countries (Wertz 1998).

Health-centred frameworks focus on the fetus. Giving birth to a child with a serious disease or condition should be avoided by means of selective abortion. The central value to be realised is to give birth to a healthy child. A sub-category consists of public health frameworks that strive for a reduction of the incidence of serious diseases and conditions in the population. In case of a test result showing that the fetus has a serious medical condition, health professionals should – explicitly or implicitly – advise the woman to terminate. Wertz has found that what I here call health-centred frameworks are rather generally adopted in developing countries in Asia – in particular China – and in many former communist countries in Eastern Europe (Wertz 1998; Mao and Wertz 1997).

Anti-abortion frameworks and restrictive abortion frameworks also focus on the fetus, but in distinction to health-centred frameworks selective abortion is considered morally wrong or at least wrong unless certain very strict medical conditions are met. Health professionals should recommend women not to terminate the pregnancy on the basis of test results or at least avoid giving information in a pessimistic (negative) way. As pointed out by Wertz, a tendency to avoid pessimistic counselling can be found in, e.g., Catholic countries such as Poland and many Latin American countries (Wertz 1998).

Within these policy frameworks, more specific goals for prenatal genetic counselling are prescribed. At least four different possible goals have been proposed (cf. Nordgren 2001, 2002, 2008):

- to give adequate information about prenatal tests and test results,
- to give assistance in decision-making,
- to give psycho-social support, and
- to avoid giving advice regarding the content of a decision (non-directiveness).

The last one of these possible goals – non-directiveness – is perhaps the most controversial. It is accepted within client-centred frameworks (perhaps with an exception if the client explicitly asks for advice), but not in health-centred or anti-abortion frameworks. The goal of giving adequate information is uncontroversial and part of all three types of frameworks. The goal of giving assistance and the goal of giving psycho-social support are more controversial and may or may not be included in all three types of frameworks. However, the more precise meaning of each possible goal may be interpreted in different ways, and there may be differences in emphasis.

8.3 The Normative Significance of the Empirical Findings

Now, do the empirical findings of the EDIG project say anything about which of these policy frameworks *should* be adopted? No, not in and by themselves, but given certain ethical and other presumptions they may say something significant. A metaphor might illustrate this. Proponents of different policy frameworks look at the empirical findings through differently coloured glasses. Given certain coloured glasses – ethical and other presumptions – the results may say something important. Given others, they don't.

However, the empirical findings may also be significant in another more indirect way. They may show that certain assumptions about the attitudes and views of women undergoing prenatal testing are inadequate, and thereby undermine the policy that these assumptions are used to support.

Let me first discuss how ethical and other presumptions may determine the significance of the empirical findings of the EDIG project. I will analyse the significance in terms of four different aspects:

– relevance,
– weight,
– interpretation, and
– application.

8.4 Empirical Finding I: Women Tend to Prefer that the Decisions are Left to Them

The first empirical finding is that most women want the decision to be left to them, although some women want some advice. Does this finding suggest that we *should* adopt a policy of not recommending women to make a particular decision but leave the decision to them?

8.4.1 Relevance

Adherents of client-centred frameworks would find the result clearly relevant. The primary goal of prenatal genetic counselling is to satisfy women's preferences and/ or needs. This view may be based on different ethical presumptions such as the principle of respect for autonomy, the principle of strengthening autonomy, the principle of beneficence, or utilitarian preference-satisfaction. Due to ethical presumptions like these, empirical knowledge about women's preferences and needs becomes ethically relevant. If the majority of women prefer that the decisions are left to them, then they should be left to them. But the results also show that some women may want some advice, and this should also be taken seriously. What this means more precisely may vary from one version of the client-centred frameworks to another.

For proponents of health-centred frameworks, the finding is largely irrelevant. What counts is a healthy child (according to individualistic frameworks) or the reduction of the incidence of serious medical conditions in society (according to public health frameworks), not women's preferences. The ethical presumption behind this view could be the principle of non-maleficence understood in terms of minimising suffering and low quality of life. Given this ethical presumption, women should be recommended – explicitly or implicitly – to terminate the pregnancy when a test result shows that the fetus has a serious medical condition. However, the empirical finding is of some indirect relevance. It is not completely irrelevant. In order to be able to implement the goal of prenatal genetic counselling – avoiding the birth of a child with a serious medical condition – in a respectful manner, you have to know how women view these things. The knowledge provided by the EDIG study may inform you about this. Moreover, you can refer to the finding that some women in fact do prefer to get some advice and point out that you want to meet this wish.

Supporters of anti-abortion frameworks would also find the empirical result largely irrelevant. What counts is that abortion is morally wrong, not women's preferences. The ethical presumption behind this view is an ethical position stressing the full moral status of the fetus. This position can be supported by a principle of human dignity or sanctity of life. Given this ethical presumption, women should be recommended – explicitly or implicitly – not to terminate their pregnancy due to the test results. However, as in health-centred frameworks, the finding may be of some indirect relevance. If the goal of prenatal genetic counselling – avoiding selective abortion – is to be successfully implemented, it is important to know at least something about women's views and preferences, so that one can approach them in a sensitive way. As in health-centred frameworks, one can also refer to the empirical result that some women want some advice and argue that one wants to satisfy this preference.

8.4.2 Weight (Within a Client-Centred Framework)

Thus, from the perspective of client-centred frameworks the empirical finding is clearly and directly relevant, but from the perspectives of the other types of frameworks it is only of indirect relevance. In discussing the finding, I will therefore from now on only focus on client-centred frameworks. However, even within this category there might still be different opinions regarding the weight of the finding. How important is it? The assessment depends on which ethical principles are considered the most important.

Before I clarify this, let me briefly describe three types of models that can be found within the category of client-centred frameworks:

- information models (cf. Ad hoc Committee on Genetic counseling 1975; Baumiller et al. 1996),
- assistance models (cf. Shiloh 1996; Kessler 1997), and
- psycho-social models (Weil 2003).

These types of models differ in how different goals are emphasised and interpreted (see above). In information models, the key goal is to provide as adequate and neutral information as possible about tests and test results, and then leave the decision to the women. According to assistance models, health professionals should challenge the views of the women in order to help them discover their true value commitments and reach a well-considered decision. Psycho-social models stress the unique needs of each woman in her particular social and cultural context: some women may need special information, some assistance in decision-making, and some psycho-social support because they experience distress, depression, anxiety, or guilt feelings (for a more extensive analysis, see Nordgren 2008; cf. Weil 2003).

Now, if you adopt an information model, emphasising the principle of respect for autonomy, you would find the result that women tend to prefer to have the decisions left to them very important, and you may stress the importance of giving adequate and neutral information in order to give women the opportunity to decide for themselves.

If you embrace an assistance model, based on the principle of strengthening autonomy, you would certainly consider the finding – that women tend to prefer to have the decisions left to them – as carrying some weight, but you would also stress the other aspect of the finding, namely that some women want some advice in decision-making. This would be in line with emphasising the importance of not only giving adequate information but also challenging the women in order to help them make a well-considered decision true to their own basic value commitments.

If you support a psycho-social model, stressing the principle of beneficence, both aspects of the empirical finding – that women tend to prefer to have the decisions left to them and that some women want some advice in decision-making – would be considered somewhat less important, because giving too much attention to this finding might suggest too strong an emphasis on the intellectual aspects of decision-making. You would rather stress its psycho-social aspects and find support for this in the other major empirical finding of the EDIG project mentioned above, namely that distress is high when women make a decision to terminate their pregnancy.

8.4.3   Interpretation (Within a Client-Centred Framework)

For those who consider the empirical finding relevant and ascribe it at least some weight, the question arises how to interpret it more precisely. One option is to interpret the finding in terms of needs that should be met by the health professionals. Those in favour of information models would then stress that in order to make their own decisions women need neutral – and adequate – information. Proponents of assistance models would focus on women's intellectual needs. In order to make their own well-considered decisions true to their basic value-commitments some women need intellectual assistance. Adherents to psycho-social models would rather emphasise the emotional needs of women in distress.

8.4.4 Application (Within a Client-Centred Framework)

How should the finding be applied in terms of supporting a particular professional policy of prenatal genetic counselling? Common to all three models is an application in terms of a policy of non-directiveness. In order to satisfy the clients' preferences and/or needs, no recommendation should be given regarding the content of decision, i.e., whether or not to have the test or whether or not to terminate the pregnancy on the basis of test results. The only exception might possibly be if the woman explicitly asks for advice.

However, as Biesecker stresses, the notion of non-directiveness can be understood in many different ways (Biesecker 1998). Actually, the three different models represent three different views on the policy of non-directiveness. Proponents of information models would suggest a policy of giving as adequate and neutral information as possible. Those in favour of assistance models would propose a policy of challenging the expressed values of the women in order to help them make well-considered decisions true to their own basic value commitments. Supporters of psycho-social models would accept a policy of non-directiveness in the sense of not recommending a particular decision, but would also point out that this does not rule out giving psycho-social support. Actually, such support is even more important in many cases.

8.5 Empirical Finding II: Distress in Decision-Making and Afterwards

What about the second empirical finding that women experience distress when they make a decision to terminate their pregnancy and for some time afterwards? Is this finding relevant, and if so what weight should be attributed to it, how should it be interpreted, and how should it be applied?

8.5.1 Relevance and Weight

In distinction to the first major empirical finding, this finding is possibly – but not necessarily – relevant to all frameworks. Its relevance depends on whether or not the goal of giving psycho-social support is included in the framework. This in turn depends on whether or not the ethical principle of beneficence – in one interpretation or another – is considered important. Moreover, there might be differences in the weight attributed to the finding, and this depends on the relative weight of the goal of giving psycho-social support compared to other goals, which in turn depends on the relative weight of different ethical principles.

Proponents of client-centred frameworks could perhaps be expected to find this finding relevant, because of their focus on the clients' preferences and/or needs.

However, even within this category of frameworks there may be different views in this regard depending on the more precise ethical presumptions.

If the goal of giving psycho-social support is included in the policy, which might not be the case, supporters of information models would find the empirical result relevant but carrying only minor weight. Stressing the principle of respect for autonomy, the main goal is to provide adequate and neutral information, not to give psychological support to women experiencing distress, and definitely not for a long time after the termination of the pregnancy. This would be a task for psychologists, but hardly for genetic counsellors.

Defenders of assistance models might reason in a similar way. With reference to the principle of strengthening autonomy, they would maintain that the key task is to give intellectual assistance in decision-making, not professional emotional support in a crisis.

Proponents of psycho-social models would naturally find the empirical finding relevant, but would in distinction to proponents of the two previous types of models stress its weight. The principle of beneficence, which is the basic ethical presumption of this type of approach, would imply that it is a highly significant fact that women experience emotional distress, not only in decision-making but in some cases also afterwards.

Adherents of health-centred frameworks and supporters of anti-abortion frameworks, respectively, could find the finding relevant, but not necessarily so. This depends on whether or not the goal of giving psycho-social support is included in their respective frameworks. Moreover, if they find it relevant, they might still differ regarding the attributed weight.

8.5.2 Interpretation

If the empirical finding – due to certain ethical presumptions – is considered relevant and carries at least some weight, the question arises how to interpret it more precisely. Given the finding, how are the needs of women to be understood? Regardless of how much weight is attributed to the finding, proponents of all frameworks that include the goal of giving psycho-social support – whether client-centred, health-centred or anti-abortion – and who therefore want to apply it in terms of supporting a particular professional policy, would need to interpret it before doing so. To interpret the finding is to put it into a broader framework of psychological and/or social presumptions.

The psycho-social needs of the women can be interpreted in different ways. Actually, there are two possible components: psychological needs and social needs. It is possible to focus on psychological needs only, on social needs only, or on both psychological and social needs.

The psychological needs may be interpreted in different ways depending on which psychological approach one adopts. Let me give a few examples. A cognitive behaviourist might focus on how the stressful situation is conceptualised and point out the need for alternative conceptual frameworks and new ways of problem-solving in order to reduce the stress (cf. Biesecker 1998). A proponent of humanistic psychology might emphasise women's need to use their positive abilities to take control of their distress and restore a harmonious state of mind, i.e., feel good (cf. Biesecker 1998). A psychoanalyst might highlight unresolved unconscious conflicts and ambivalences rather than reducing distress and suffering. They might stress that what the women need is self-knowledge, and obtaining such knowledge may include some psychological pain. The goal is not simply to feel good (cf. the Chap. 11 in this book). There might also be differences between different approaches regarding how much psychological support is needed in terms of counselling and for how long after the decision.

Similarly, the social needs may be interpreted differently depending on which social approach one presumes and also depending on the social system of the particular country. If a woman deliberates on giving birth to a disabled child she might fear the economic burden of raising the child in absence of sufficient support from society. This need for support is a very real problem in many developing countries and also in some developed countries (cf. Wertz 1998). There might also be other less dramatic social needs such as the need to meet other women in a similar situation. A woman might feel lonely and need a social network.

8.5.3 Application

Given these various interpretations of women's needs, different methods to meet the needs may be suggested. With regard to meeting the psychological needs, cognitive behaviourists, humanistic psychologists, and psychoanalysts would provide different approaches. And the methods of meeting the social needs may also vary. In a society with a developed welfare system, the counsellor might present the options of social and economic support to parents with a disabled child. The counsellor might also provide an opportunity for women to meet other women who have already given birth to disabled children, for example, children with Down syndrome. In this way they may get a more realistic picture of how such a situation might look like. Women may also be introduced to groups of women to get support and learn that they are not alone with their problems.

8.6 Empirical Findings May Undermine Normative Positions

So far I have focused on how the different ethical presumptions of different policy frameworks determine the significance attributed to the empirical findings. However, empirical findings may also be significant in another more indirect way. They may show the inadequacy of assumptions about the psycho-social situation, attitudes and views of women undergoing prenatal screening and diagnosis.

This may not necessarily show that a particular policy framework is ethically wrong, but by showing that its empirical support is inadequate it undermines the policy and gives its proponents reason to reconsider their view. Let me give a few examples (note that these are illustrative and hypothetical examples only, and that some are less realistic than others).

Take for example somebody who believes that all or at least most women considering prenatal genetic testing want recommendations from the health professionals whether or not to have a test or whether or not to terminate the pregnancy due to the test results, and who on the basis of this assumption argues for a policy according to which women should be given such recommendations. This assumption about facts would be shown to be inadequate by the empirical finding that most women want the decision to be left to them. This would in turn undermine the policy and suggest that it should be reconsidered.

Take another person who thinks that women considering prenatal genetic testing do not want recommendations regarding a decision but prefer to decide for themselves, and who with reference to this assumption proposes the policy that women should never be given any recommendations. This empirical assumption would be inadequate given the finding that some women actually do want some advice, and this would undermine the policy proposal.

Moreover, imagine somebody who believes that for all or at least most women it is not particularly distressing to undergo prenatal genetic testing and who therefore suggests a policy that does not include psycho-social support as a goal. This belief – or assumption – would be shown to be inadequate by the empirical finding that distress is high at least when women make a decision to terminate their pregnancy. This would in turn undermine the policy to the extent that it is based on a wrong assumption.

Finally, take somebody who assumes that the distress some women feel when they make a decision to terminate their pregnancy vanishes rather quickly, and who therefore suggests a policy according to which psycho-social support needs not be provided for some time afterwards. This assumption would be shown to be inadequate by the empirical finding that some women actually do remain distressed for some time after the implementation of their decision. This would undermine the policy proposal.

It is vital to note, however, that in neither of these cases the empirical finding would show that the policy is wrong. The finding only undermines the policy in the sense that its factual support is shown to be inadequate. This suggests that the proponent of the policy either looks for other types of support or changes the policy.

8.7 Conclusion

The conclusion of this investigation is not that it is impossible to use the empirical findings of the EDIG project as support for a particular policy of prenatal genetic counselling. The point is rather that the findings do not speak for themselves.

Their significance depends on already accepted ethical and other presumptions. However, the findings may also show that certain assumptions about women's attitudes and views are inadequate. Thereby they may undermine the policy that these factual assumptions are used to support. This may give proponents of the policy reason to re-evaluate their view.

References

American Society of Human Genetics (ASHG) Ad hoc Committee on Genetic counseling (1975) Genetic counseling. Am J Hum Genet 27(2):240–242
Baumiller RC, Cunningham G, Fischer N, Henderson M, Lebel R, McGrath G, Pelias MZ, Porter I, Roper WN (1996) Code of ethical principles for genetics professionals: an explication. Am J Med Genet 65(3):179–183
Biesecker BB (1998) Future directions in genetic counseling: practical and ethical considerations. Kennedy Inst Ethics J 8(2):145–160
Fischmann T, Pfennig N, Läzer KL, Rüger B, Tzivoni Y, Vassilopoulou V, Ladopoulou K, Bianchi I, Fiandaca D, Sarchi F (2008) Empirical data evaluation on EDIG (Ethical Dilemmas due to Prenatal and Genetic Diagnostics). In: Leuzinger-Bohleber M, Engels E-M, Tsiantis J (eds.) The Janus face of prenatal diagnostics: a European study bridging ethics, psychoanalysis, and medicine. Karnac, London, pp 89–135
Kessler S (1997) Psychological aspects of genetic counseling. XI. Nondirectiveness revisited. Am J Med Genet 72:164–171
Mao X, Wertz DC (1997) China's genetic services providers' attitudes towards several ethical issues: a cross-cultural study. Clin Genet 52:100–109
Nordgren A (2001) Responsible genetics: the moral responsibility of geneticists for the consequences of human genetics research, vol 70, Philosophy and medicine. Kluwer Academic, Dordrecht
Nordgren A (2002) Wisdom, casuistry, and the goal of reproductive counseling. Med Health Care Philos 5:281–289
Nordgren A (2008) Prenatal genetic counselling: conceptual and ethical issues. In: Leuzinger-Bohleber M, Engels E-M, Tsiantis J (eds.) The Janus face of prenatal diagnostics: a European study bridging ethics, psychoanalysis, and medicine. Karnac, London, pp 307–325
Shiloh S (1996) Decision-making in the context of genetic risk. In: Marteau T, Richards M (eds.) The troubled helix: social and psychological implications of the new human genetics. Cambridge University Press, Cambridge, pp 82–103
Weil J (2003) Psychosocial genetic counseling in the post-nondirective era: a point of view. J Genet Couns 12(3):199–211
Wertz D (1998) Eugenics is alive and well: a survey of genetic professionals around the world. Sci Context 11(3–4):493–510

Chapter 9
Taking Risk in Striving for Certainty. Discrepancies in the Moral Deliberations of Counsellors and Pregnant Women Undergoing PND

László Kovács

Abstract Analyzing arguments in decisions about PND shows that the ethical concern of health care professionals differs considerably from the ethical concerns of pregnant women. Health care professionals are more concerned with the right interpretation of medical facts and the application of established principles of counselling. Pregnant women on the contrary focus on broader life circumstances, relations in the family and their own emotions. Both ethical perspectives are limited but can usefully complement each other. However, in certain cases there can be conflicts between them. In these cases it is not enough to refer to scientific authorities, the principle of non-directiveness or to the principle of autonomy of the pregnant women. Ethically sound solutions have to integrate both perspectives and evaluate all facts even at the costs of some less central goals of PND.

Keywords Decision-making • Genetic counselling • Pregnant women • Prenatal genetic diagnosis

9.1 Introduction

Prenatal genetic diagnosis (PND) was introduced in most European countries during the 1970s. The initial goal of this invasive technique was to provide couples who had a high probability of giving birth to a child with serious disabilities with a definite diagnosis. This diagnosis was to help them in making decisions about the continuation or termination of the pregnancy. Since then the technique has been established more widely within obstetrics and gynaecology and is offered to almost all pregnant women who are at risk.[1] To test under certain circumstances has

---

[1] This is especially true for Germany where 70–80% of all pregnancies fulfil the criteria for "high risk pregnancies". In most of these cases an invasive PND is offered (Lux 2005, p. 19). This leads to invasive diagnostics in ca. one of six pregnancies (BZgA 2006, p. 33).

L. Kovács (✉)
University of Tübingen, Wilhelmstr. 19, 72074 Tübingen, Germany
e-mail: laszlo.kovacs@uni-tuebingen.de

become routine for many pregnant women. Routine means that pregnant women undergo some diagnostic tests as an integral part of pregnancy. Routines of this sort are ethically questionable since they concentrate on some aspect of a dilemma and disregard other aspects which may be ethically relevant. This extended application of PND indicates complex ethical considerations with respect to the harms and benefits of this medical technology regardless of the subsequent decisions about termination of the pregnancy after a positive diagnosis. The decision to undergo PND is associated with different ethical perspectives: those of the health care professionals and the pregnant women. Therefore I focus on the ethical justification of the decision to undergo PND from these two different perspectives. The first part of the text will scrutinize inherent ethical difficulties of providing PND. Firstly I present crucial ethical principles and medical facts relevant to ethical analysis and secondly I show the consequences of including them in counselling and taking seriously the ethical principles. In the second part I will summarise and interpret some of the answers from an empirical survey highlighting the moral justification patterns of pregnant women. I conclude with some inherent differences between the two ways of moral reasoning and ask for an integration of both aspects as part of pre-test counselling.

9.2 Development of Counselling Strategies Concerning PND

Through the development of modern medical technology new prenatal diagnostic methods were introduced in the second half of the twentieth century. In connection with abortion laws of the 1960s and 1970s PND opened up the possibility of reproductive decisions about genetic conditions of the fetus. Beyond the technological achievements and the modification of the legal framework, a third conceptual change occurred in prenatal medicine: the concept of non-directiveness became more and more accepted.[2] Physicians adopted a non-paternalistic communication style: counselling. The concept of counselling has also developed during the last decades.

In the 1970s non-directive counselling implied that the physician summarised the family history with the couple, performed tests, presented medical risks and, at the end, s/he gave advice to the future parents. Mahn (1979) presented this concept impressively by identifying five types of advice: (1) It is recommended to have children without any restriction (2) it is recommended to have children but only after amniocentesis, (3) it is recommended to have children but not with every partner, (4) it is recommended to have children but only with certain restrictions,

---

[2] The inventor of the concept in genetic counselling, the American geneticist Sheldon Reed refers to the findings of Carl Rogers who elaborated effective methods for a client centred therapy and analysed the relationship between counsellors and counselees in psychotherapy (Rogers 1942; cf. Porter 1977).

(5) it is not recommended to have children.[3] The justification for the advice was based on medical rationality and the counsellor had to decide what kinds of disorders he considered serious enough to be prevented. Parents were, nevertheless, free not to follow the advice. This style of counselling was important for the two central goals of counselling: (1) prevention of the birth of disabled children (a goal presumed to be beneficial for parents and society) and (2) help in overcoming the unnecessary fears of parents of having disabled children (Mahn 1979, pp. 92–96). The second central goal was emphasized as a kind of utilitarian psychological reason since almost all parents could be reassured and only a very small number of high risk parents had to be let down (PND had psychologically beneficial effects). Still, this last group had the freedom to take the risk of having children even if it was recommended not to do so (Cremer et al. 1983, pp. 1–2).

In the following years reflection about the normative dimensions of medical facts led to the awareness that it is not possible to derive moral decisions from medical facts alone (medical indication) without normative judgements. Hence the counselling strategy changed towards an even less directive model: The counsellor was now presenting facts relevant for a decision but at the same time he had to be absolutely reserved concerning the decision. This counselling attitude was considered beneficial mainly for ethical reasons. It corresponded to the observed moral pluralism of post-modern Western societies where physicians have no appropriate competences to judge the moral preferences of autonomous patients. Health care professionals were therefore not allowed to presume a certain moral choice for pregnant women in a dilemma. This principle of non-directiveness matched the dominating individual paradigm in ethics claiming that the state as well as the physician has to withdraw from the paternalistic role (Jungermann et al. 1981).

This kind of non-directiveness shaped counselling so that moral duties of counsellors and their competences in morally relevant aspects of the decision became obsolete but counsellors were entitled to provide the factual fundaments of normative judgements (Kovács 2008). Initially this change may seem marginal but it is actually central. It means that public interests in health or the economical use of public resources have no place in the decision. Two principles became central: the non-directiveness of the counsellor and the autonomy of the pregnant woman. The first goal of counselling (public health) was formally abandoned. The second central goal of PND, i.e. the presumed psychological effect, has not been challenged by the principle of non-directiveness. Pregnant women have the right to make the final decision whether they make use of PND and counsellors are invited to empower pregnant women in their decision-making – including the ethical consequences of the decision.

---

[3] For Germany cf. Mahn (1979), similar to the style of counselling in Western Europe, counsellors in America (cf. Lubs and de la Cruz 1977) and in Eastern European countries (cf. Czeizel 1981) used comparable criteria.

9.3 Empowering for Decision-Making

In order to aid decision-making health care professionals feel their responsibility is to provide appropriate information about PND. Providing all relevant and correct information is part of their professional and legal duty. Thus they are very concerned to find the best way of doing this and determine the most relevant and accurate information to enhance communication.

However, it is difficult to define what is appropriate to communicate in each case. Furthermore, comprehensive and neutral information is a concept that has several limitations and inherent difficulties. Telling "the whole truth" including all relevant medical and technical information, as well as professional and individual moral presuppositions, is of course not possible. To achieve informed consent counselling must focus on certain substantial aspects: (1) what tests are available or offered, (2) what outcomes they may produce, and (3) how to interpret the outcomes in order to come to a decision. Providing this information is a result of personal and professional choices involving several normative factors.

(Ad1) Prenatal genetic tests can be divided in two categories: invasive (like amniocentesis) and non-invasive ones (like bio-chemical screening – the so called triple test). Invasive tests have a better predictability than non-invasive ones but they are associated with a higher risk to the fetus and it usually takes longer to obtain the results. Because of the higher associated risk, invasive tests are usually offered only to pregnant women at higher risk than the test itself causes. Behind the decision about the level of risk at which invasive tests are offered to pregnant women there are normative judgements by the profession. Some technically possible tests are explicitly excluded or implicitly ruled out by professional guidelines whilst others are offered as routine like ultrasound screening. Medical guidelines define what kinds of disorders are "reasonable" to screen for. Such value judgements are subject to professional mainstreams. A well known example is the fact that many countries in Europe offer amniocentesis to women above the age of 35. They justify this by balancing the potential risk of miscarriage compared to the risk of having a baby with Down syndrome. The invasive examination has a risk of pregnancy loss of about 1:200 which is as high as the occurrence of trisomy 21 at the age of 35 years (risk at 35 is usually given as 1 in 365). This routine offer is based on a hidden moral presumption: trisomy 21 is equivalent to the loss of pregnancy. This threshold appears to be based on medical criteria but it is not the case. In France and Norway the same threshold is the age of 38, in Finland at 39 (Boyd et al. 2008, p. 692). These differences are not related to the lower frequency of Down syndrome in these countries but to the normative judgement about the value of a life with Down syndrome related to the risk and the costs of the test – this is the age where risk from miscarriage and risk of Down syndrome correspond. Who is offered what kind of test is partially based on normative assumptions that are not open to discussion in counselling. Health care professionals are or should be aware of the fact that they are not free of professional norms, even if they do not bring up these issues in every counselling session.

(Ad2) The medical information produced by PND is so complex that it regularly transcends the capacity of non-professionals. For a visual summary of the morally relevant aspects of a test result a "prediction cube" is presented below that illustrates the four most relevant medical dimensions of the information: expected symptoms, severity of the symptoms, probability of the severe and less severe symptoms and the time when the symptoms manifest themselves.

Symptoms: A genetic test will identify a specific genetic condition but not all possible genetic disorders. It is important to differentiate between tested genetic conditions (genotype) and symptoms (phenotype). Some of the genetic conditions cause severe disability or even early death; others are associated with rather mild symptoms or symptoms that pregnant women would consider as bearable for themselves and the child. It has to be emphasized that one single genetic condition may lead to very different symptoms. Thus a detected gene or a chromosomal disorder does not give definite information about the life expectancy and the quality of life. Rather there is a list of probable but not inevitable consequences. For a mother-to-be the detected genes should be secondary and the symptoms should be the focus of decision-making: Can she manage living with a child with the predicted symptoms? Decisions concerning certain symptoms can be made before the test is carried out: If the pregnant woman does not consider a certain kind of disability (e.g. a gene for Huntington's disease or Down syndrome) as decisive for further actions like the termination of her pregnancy,[4] the test may be contra indicated since the test is associated with risks to the fetus. The International Huntington Association demands a statement from the pregnant woman before PND that she is going to terminate the pregnancy if the test result is positive. Otherwise they see no valid reason to test because a test result would compromise the right of the child not to know his genetic status (IHA 2008, cf.: http://www.huntington-assoc.com/). This example shows that the kind of expected symptoms may be relevant for the decision whether to undergo PND. Offering a diagnostic measure that puts the life of the fetus at risk must be justified by sufficient reasons.

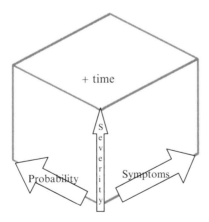

---

[4] Most conspicuous test results serve selective abortion in Europe (for Huntington's disease see: Decruyenaere et al. 2007, p. 454, for the more frequent Down syndrome and neural tube defects see Boyd et al. 2008, p. 693).

Probability: Not every test gives certainty. Having a positive diagnosis about one single disorder is a matter of fact but it needs interpretation. The counsellor has to declare how probable it is that the finding is correct.[5] The issue of probability is important for appropriate moral deliberation. Individual risk perception differs from real risks. Positive tests with low probability impress people too much. Studies show that making one possible risk concrete out of a group of unqualified risks leads to a judgmental bias and causes the overestimation of the named risk.[6] The offer of a prenatal test creates anxiety about the nominated disorder in contrast to non-nominated ones. Therefore many counsellors argue not to tell pregnant women about every possible disorder. Instead they present only the most frequent disorders (a matter of choice) and hope to avoid unnecessary additional anxiety during pregnancy. In this respect it is difficult to avoid being somewhat paternalistic.

A second problem with the communication of probabilities is that most of the people are not familiar with the interpretation of statistical values. It has been shown that even the framing of probabilities leads to different decisions (Wüstner 2001). The small difference in presenting the probabilities in absolute or in relative terms influences patients' preferences.[7] In this regard, Bramwell et al. (2006) surveyed the capacity of different stakeholders in prenatal diagnostics (pregnant women, their partners, midwifes and obstetricians) on their interpretation of probabilistic information. They presented the probabilities of one genetic test to two mixed groups of stakeholders: either in percentages (1%) or in frequencies (1 in 100). There was a statistical difference between the two groups in understanding the probability of the test correctly: Frequencies were somewhat better understood. Although obstetricians performed slightly better than the rest, almost 86% of the responses were incorrect. Very low percentages were more frequently overestimated. Many professionals were confident about their responses – albeit incorrect (Bramwell et al. 2006). A further difficulty with the communication of probabilities is in interpreting the correct statistical information for lay people. Such communication difficulties appear in several fields of translation of professional knowledge into public language (Kovács and Frewer 2009). The following quote from an EDIG questionnaire is quite representative of this problem:

> I found it very frightening that due to the triple-test my risk for trisomy 21 is at 1:60 instead of 1:1,780. The last one would be in accordance with my age. After I translated 1:60 into the likelihood of 1.6% for trisomy 21 compared to a likelihood of 98.4% for having a healthy baby it was not so frightening... (translation: LK)

---

[5] Many couples tend to have high expectations towards the validity of prenatal tests although some of them have a considerable amount of false positive results. Invasive genetic tests like amniocentesis gives almost perfect certainty. Conspicuous ultrasound screening results can usually be clarified by further invasive tests.

[6] Slovic et al. called this phenomenon of lower perceived risks of not nominated risks "out of sight out of mind" (Slovic 1982, p. 470 f.).

[7] Malenka et al. presented the benefits of one single medication to patients in two different ways: in relative terms and in absolute terms. "56.8% of patients chose the medication whose benefit was in relative terms. 14.7% chose the medication whose benefit was in absolute terms" (Malenka et al. 1993).

Severity: Even identical genetic disorders differ in phenotype and severity of the symptoms. The possibility of prediction is very limited for the majority of the tested genetic disorders. Estimations often allow only a description of tendencies or best and worst case scenarios. For reasons of correctness after a positive test result, an explanation of the concrete problems caused by the detected genetic disorder must be given. For example, in Myotonic Dystrophy the expected severity is related to the number of trinucleotide repeats in the gene. It is probable that fetuses with a positive diagnosis (between 100 and 1,000 repeats) will develop to adulthood having amyosthenia (mainly in the legs), certain heart disorders, cataracts and other problems. But which symptoms will appear and how severe they will be is hard to predict. The same is true for symptoms of Down syndrome and even for the paradigm of predictable genetic diseases such as Huntington disease. By presenting the likely symptoms it may be necessary to detect and correct extreme expectations or identify biases based on single case experiences.[8]

Time: Many diagnoses include a perspective about the timing of the development of symptoms during life. For the decision to have a baby or terminate the pregnancy, it may make a difference if the symptoms develop early or late in life. Some of the early onset disorders can be treated so well in the first years of life that they do not affect the later development of the child. Other genetic disorders do not cause symptoms until adolescence or even until the age of 30 or 50. For babies born now with such a condition, there is an opportunity for medicine to find an adequate prevention or treatment of the symptoms and to change the perception of the disease.[9]

(Ad3) Last but not least counselling has to address basic questions about what results mean in a broader context. In this respect an important task of counselling is to point out and to challenge the reductionist perspective of medical technology. Counsellors have to present medical knowledge but they also have to create discursive freedom from medical facts. They have to present test results as concrete as possible and create a distance between medical information and the complexity of life decisions. This requires a high level of reflection and is probably the most difficult to achieve.

Test results stay in the realm of important medical knowledge and attract attention. They demand action, more technology and more medical control. They change the ignorance into predictive and hence uncertain knowledge. They claim to make uncertainties more certain by using even more technology. They promise the possibility of eliminating certain kinds of suffering and so they claim to be implemented regardless of other kinds of suffering. They change the unstructured time of the

---

[8] Available experiences have a great influence on the heuristic in which a person evaluates the frequency or the severity of the diagnosis (cf. Tversky and Kahneman 1973). Leuzinger-Bohleber et al. present the psychological complexity of a case where the decision to terminate the pregnancy is based on an inaccurate perception of the severity of a genetic disorder the pregnant women experienced in her family (Leuzinger-Bohleber et al. 2008a, pp. 188–195).

[9] E.g. Familial Adenomatous Polyposis is a hereditary genetic disorder leading to colon cancer but now there is a possibility of prevention. So having a genetic disorder could be uncoupled from symptoms.

pregnancy into a structured time of possible interventions. Therefore technology has been identified by Wieser (2006) as an "individual actor" who is coercive in terms that it defines substantial criteria of the action which are not discussible since you cannot discuss with technology. It is just there and it influences moral deliberation by defining the basic terms of the discussion.[10]

Normative effects of the technology and of the health care system are of course not easy to recognise in PND from the internal perspective of those involved. Medical professionals are trained to think in medical terms and to apply medical criteria. At the same time they have to support pregnant women in relating the medical perspective to their overall life conditions and to the perspective they may have with their future child. Pregnant women often do not even realise that there are imposed medical categories and that they have to manage a conflict between the realm of medical thinking and their own life. But they take basically another perspective.

9.4 Moral Reasoning of Pregnant Women

Pregnant women are involved in PND on two different levels: first they are personally and bodily involved in PND and second they are the ones who have to make decisions. Their double-role is morally ambivalent since they have to represent their own interests and the interests of their fetus. Pregnant women experience that these two interests sometimes conflict and the moral ambivalence causes difficulties in their decision-making. By deciding to have PND they hope that everything will be all right but are afraid of conspicuous results.

9.4.1 I Only Want to Have A Healthy Child

Members of the EDIG project asked pregnant women before and after undergoing PND to comment on their psychological and moral distress. Answers were analysed in an interdisciplinary setting by psychoanalysts, medical care givers, ethicists and cultural anthropologists.[11] One of the first questions of the EDIG questionnaire (before the PND was carried out) asked pregnant women "Do you have thoughts, fantasies and wishes about your developing baby?" The majority of answers emphasized only the health of the child, for example: "I only want to have a healthy child." In discourses about PND health is obviously the most important vision

---

[10] Wieser emphasizes the goal-oriented role of technology in PND. Following the sociologist Bruno Latour he compares PND with the speed bump in the traffic: By staying there and doing nothing the speed bump is influencing the behavior of car drivers (speed reduction) and is functioning as part of an action programme. Prenatal test results similarly do nothing but are just there and also function as a certain action programme (Wieser 2006, pp. 118–119).
[11] See results in Leuzinger-Bohleber et al. (Eds.) 2008b.

concerning the future child. Reassurance that the fetus is healthy is the major motivation for prenatal diagnostics.

But the health of a fetus is a more complex issue than it could be tested in PND. It is a concept with several normative implications. First of all, the health that pregnant women hope for is rather holistic in nature and includes more than medical facts and genetic conditions. Nevertheless, the growing medicalisation of pregnancy imposes new categories on pregnant women's perception of pregnancy. Their relation to the fetus is increasingly influenced by medical technology that offers pictures, movies and much health related information in technically framed medical terms. "Health" in pre-PND-questionnaires is frequently perceived as an idealistic fact that can be tested prenatally and can be separated from "not-health". PND reduces this concept to purely medical or even to genetic aspects. Moreover, this reductionist concept of health still cannot be affirmed by genetic tests as a descriptive fact but it is rather connected to a normative concept that needs individual interpretation. The prenatal discourse about the fetus, nonetheless, is rather framed in categories of "perfect" and "defect" or "abnormality" regardless of the severity of the symptoms (Rothschild 2005). However, there is no general definition of genetic health that could be affirmed. All of us have some predisposition to different genetic conditions. Nevertheless, we would say that we are healthy, being free of symptoms, despite having such genes. The health of the fetus in non-professional discourses is termed in categories other than the health of adults. Let's make it concrete with some examples. Where is the boundary of health regarding a fetus with:

- a gene for Lynch Syndrome, an inherited form of colorectal cancer with a lifetime risk of some 80%? We know that this risk can significantly be reduced by a regular screening program after the age of 45 (Komaromy 2000a);
- a gene for Familial Adenomatous Polyposis (FAP), which can lead to similar symptoms like Lynch syndrome but much earlier (starting by the age of 15) and a more frequent screening is the only secure way to prevent cancer (Komaromy 2000b);
- a gene for Huntington's Disease which regularly leads to serious symptoms between 30 and 50 years of age and finally to death unless medicine finds therapies in the next decades (cf. IHA, http://www.huntington-assoc.com/);
- a gene for dwarfism (Koch 2008);
- a gene for Duchenne Muscular Dystrophy with a prognosis of 1 or 2 years free of symptoms and some 10 years with a progressive degeneration of all muscles (Kirmse 2006);
- a gene for Zellweger syndrome with a prognosis of some weeks to months of life with deteriorating condition.

These conditions are very different in nature but in lay discourses we would consider at least some of them as "not healthy". There are no objective or intersubjective ethical principles that could help in determining a threshold of "health" of the fetus. Moral judgements about individual cases are necessarily subjective.

Concerning the actual health status, however, such fetuses are healthy fetuses. Symptoms, disability or suffering are only predicted for the future.

PND counselling in the beginning openly aimed at the overcoming of unnecessary fears of parents of having disabled children. In the actual ethical discourse about PND this aspect does not receive much attention. Unlike this change, the psychological relief is still a central issue for pregnant women in using medical technology during pregnancy. They make use of prenatal diagnostics in order to know that everything is going fine and to resolve the unbearable feeling of uncertainty. "I only want to have a healthy child" also means that pregnant women before undergoing PND do not stay in the realm of the complex normative evaluation of medical knowledge. Their focus is not so much on the critical status of the achievable knowledge rather on relations and emotions.

9.4.2 Relationality as Rationality[12]

In the questionnaires, pregnant women who gave reasons against having a child with a disability took into account their particular life circumstances and those of others involved. In summarizing the answers a new central category can be identified that has not played a central role in professional moral deliberation: relations. Facing this moral conflict, pregnant women do not see themselves as isolated rational decision-makers deliberating pros and cons of the medical information but as staying in relation to and with responsibility for relevant others.[13] Most of them report on a deliberation about the effect on other family members of a child with disabilities with regard to psychological distress, social exclusion and less time for other family members. In their answers they identify several relevant persons for whom they feel responsible and whom they want to protect[14] – especially protecting born children from a life with a disabled brother or sister. The following examples may give an impression of this kind of reasoning:

> That I don't know whether I could have a disabled child since my other son has to struggle enough because of his stutter and I didn't want him to be in 'competition' with a sick child.
>
> In the case that the child has a disability I don't want to have it in order to protect my small daughter and our own life, too.

---

[12] The following findings are results from the German part of the EDIG survey, i.e. answers from German pregnant women to open questions in an extensive questionnaire about psychological and ethical issues. Results from a quantitative analysis see also in Leuzinger-Bohleber et al. (Eds.) 2008b, in particular in Fischmann et al. 2008.

[13] The ideal of an independent individual has been criticised by several feminist ethicist, e.g. Corandi (2001).

[14] The first question was: "Have you talked to anyone about what you would do if an abnormality was detected?" followed by the open question: "What did you talk about?" This question asked the "about" and not the "why", still a high number of pregnant women presented moral justifications for their decision.

> A handicapped child does not come into consideration since we have responsibility for two other healthy boys. (Translation: LK)

Another type of answer presented the issue of responsibility towards themselves. Pregnant women did not emphasize the principle of autonomy in their decisions. They never write: "This is my choice". Instead, their answers accentuate interpretations with a more dimensional responsibility:

> I would not like to nurse a disabled child. The burden would be too great for me.
> As I already have three healthy children I would decide against a disabled child.
> I have various and to some extent chronic diseases; I know that my power is limited. A severely disabled child would ask too much of me. Furthermore we have no support from the family. (Translation: LK)

A third type of answers considers the quality of life of the future child.[15]

> I would not give birth to a disabled child due to respect for the child itself because I think that this would be a too drastic life situation.
> Due to my age and the connected risks of giving birth to a disabled child. The health of the child has highest priority for me. I do not consider a pregnancy *at any price*.
> How difficult it must be for a disabled child to manage this life. What if the mother is not there anymore? (Translation: LK)

Some responses comprised more than one pattern. Several of them refer to more than one of these responsibilities or combine different arguments. The following answers are examples:

> I said that I wouldn't have the child if a disability is diagnosed because I have two other healthy children (very active and stressful). Moreover I would psychologically and physically not be able to raise a disabled child because of previous strokes of fate.
> If there is a disability I do not want to have a child – for protection of our small daughter and of our own life. (Translation: LK)

Answers do not suggest who should be protected more than others. The "hierarchy of values" is a characteristic of women's interpretation of the moral problem of the principle of autonomy. Responsibility is much more the central term: towards family members and oneself. Responsibility depends on the quality of the relationships. It is very strong with members of the family and less so with the fetus. If relationship to the unborn child is an issue, it refers to future relations or rather to visions and expectations about future relations. These visions and expectations are expressed in narratives about capacities and characteristics of the future life of the child and the family.

---

[15] Because in the questionnaire it was not explicitly asked for a motivation the proportions do not present reliable quantitative data, but the qualitative nature of the answers is given. A quantitative study form the Netherlands aimed to demonstrate motivations for termination of pregnancy in case of Down syndrome showed similar results: 73% of mothers considered the burden too heavy for other children of the family and 64% considered this burden too heavy for themselves. This study also reports a higher proportion of pregnant women who consider the burden of the disorder for the future child too severe (45–92%) (Korenromp et al. 2007, p. 149.e3).

9.4.3 Emotions in Moral Decisions[16]

Another characteristic of the moral reasoning of pregnant women in the questionnaires is their fundamental orientation to emotions in moral decisions. "Health" is not contrasted to an expected medical diagnosis of "not-health" but to experiences and narratives of illness and disability which cause anxiety. Such emotions are interpreted as reasons why pregnant women undergo PND. PND makes the expectations of health concrete and provides a technically mediated coping strategy for anxiety. Medical technology serves in this context as an instrument of "making sure that everything is OK".

> Prenatal diagnosis is 'part of the regular precautionary programme'. It is reassuring to know that I did everything. This has a positive effect on my feelings during the pregnancy.
>
> It is comforting to have the possibility to exclude disability or to act accordingly if the result is bad.
>
> Amniocentesis can be recommended: it gives security for a relatively small risk. I also did a screening test (NT-test and hormone analysis): these I would not recommend: this causes a permanent confrontation with diagnostic results: one is positive (not in medical sense) and one is worrying. There is more insecurity and uncertainty than additional security. It is highly problematic especially for people with an unstable psychological disposition or if there was no decision made before the test was carried out about the process after the results are there. (Translation: LK)

Disability on the other hand does not remain an abstract idea raised by the medical counselling. Pregnant women report from single experiences of themselves and of others, visions and narratives. Such disability-stories in contrast to health-stories shape the perception, make the abstract idea of disability concrete and support the feeling of certainty in decision-making. The following quotes show how experiences and stories may shape the understanding of disabilities.

> In my job I have got to know two cases: two disabled children are now about 30 years old and need parental care round-the-clock. They are not able to manage their own life and are bed ridden. The parents are haggard, they constantly worry, are exhausted and don't know what will happen if they are not there anymore. These cases have influenced me very much. (Translation: LK)

Another future oriented type of narrative in the EDIG-study answers presents the anticipated emotional burden of living with a disabled child. Many pregnant women anticipate and estimate their psychological resilience as well as their emotional capacities to raise a child with a particular disability.

> If it is a baby with Down syndrome, I would prefer an abortion because of the severity of the disease – primarily mentally – which cannot be corrected. And I don't want to see my child seeing other children play and my child cannot do so and is unhappy.

---

[16] The following findings are results from the German part of the EDIG survey, i.e. answers from German pregnant women to open questions in an extensive questionnaire about psychological and ethical issues. Results from a quantitative analysis see also Leuzinger-Bohleber et al. (Eds.) 2008b, in particular Fischmann et al. 2008.

> For me it is important to know whether my child is disabled or not since I cannot imagine a life for myself with a disabled child: this would be a life-task that I personally cannot imagine for myself.
>
> I am a very serious human being who prefers security. My husband and I did not want to expose our son to a child that could have a trisomy. We both see ourselves as emotionally unable to manage such a life. Therefore amniocentesis offers a good possibility.
>
> I felt that I was coward that I performed amniocentesis and that I possibly decide against a disabled child. I think disabled people are important parts of our society. However, I am too much a coward to give birth to a disabled child. (Translation: LK)

Many answers show that emotions considerably influence decision-making even if they are not directly translated into decisions. The emotional effects of the outcome as well as the procedure are in some way included in the decision-making. They may be ambiguous but they are authentic and take personal circumstances into account.

9.5 Conflicts Between the Different Kinds of Moral Reasoning

The moral deliberation of pregnant women differs from that of health care professionals regarding both content and procedure. Health care professionals respect the ethical and legal framework. Therefore they focus on standards of counselling following mainly evidence based facts, validated statistical values about the actual and expected medical conditions. Non-directive counselling therefore has a hypothetical character that has to be translated into action by those who are concerned and who have to make decisions at the same time. Nevertheless, in some respects professional counselling continues to be paternalistic, e.g. in providing or withholding information or in not-offering tests for certain conditions.

Pregnant women receive information and incorporate it into a decision-making procedure which is oriented towards relational and emotional balance. Decisions about PND need, of course, to be made in a life context with sensitive evaluation of the resources and capacities of all involved. It has a great merit that pregnant women worry about relations and feelings. Answers show that they remember experiences from the past and develop visions for the future in order to translate their knowledge into concrete representation. Their perspective is individualised and their ethical deliberation cannot be reproduced by using generalizable ethical principles. A theoretical defence of this ethical position can be found by the feminist ethicist Carol Gilligan (Gilligan 1984). Care ethics, following the position of Gilligan, treats personal experiences as the most reliable source of ethics, as a special view of moral deliberation which does not need to be corrected by principles of traditional ethics but exists in its own right. Consequently, care ethics could refer to two determinants of moral consideration of pregnant women: relations and emotions. These two determinants are able to define the most important issues of the ethical discourse regardless of abstract principles. This ethical position conflicts with the less individualized, fragmented, principle based professional ethics at least in three respects.

First, offering prenatal tests on a large scale has been criticized for producing effects on social perception of disability like screening for a certain disability (Rothschild 2005). Pregnant women are so much concerned by their individual situation that they ignore morally problematic tendencies on higher levels than the individual case. As individuals, pregnant women are not able to take a critical position against social tendencies. Health care professionals, as individuals or as members of organisations, may recognise ethical problems and go against certain tendencies for ethical reasons. By integrating PND into a health care system, "action programmes" (Wieser 2006) develop that create their own internal rationality and it becomes difficult not to follow the programmes by entering the system. In this sense PND as a technology and a social practice has normative effects. As soon as there is a conspicuous finding, the examination changes its meaning. It is not a control for "that everything is alright" but the appeal for personal responsibility of the pregnant woman offering her further steps of clarification and a justification for aborting the fetus. If PND is offered, pregnant women cannot escape making a decision to undergo or not to undergo it and for being responsible for their choices (Engels 2008, p. 255). In this way routines emerge that are ethically problematic. Therefore professional guidelines try to limit the free choice of pregnant women and fall into conflict in the individual case. Criticisms of a distant position play an important role but have been criticized for being not involved in the evaluation of the individual circumstances.

Second, compared to professional ethical standards the value of a subjective narrative is questionable. A case in the EDIG study illustrates this problem. In a detailed interview Mrs. F. told the story of her "heavy, burdened childhood" with her brother who was affected by haemophilia, an inherited disorder. When she became pregnant she decided she did not want to have a child with such a future and had PND. She constructed a vision of her unborn child based on her personal experience of her brother. She anticipated the emotional strain of the disorder and, after receiving a positive test result, she decided to terminate the pregnancy.[17] From the perspective of care ethics this decision seems to be acceptable: Mrs. F.'s experience with haemophilia shapes her relation to her child carrying the gene for haemophilia. She might suffer from this vision and would cause unnecessary suffering to her child because of her relation to the illness. She is the one who can estimate her capacities and consequently she is the one who must be entitled to make the moral decision.

Referring to this narratively interpreted experience is, however, reducing ethics to the subjective estimation of the situation without available objective criteria. Furthermore, single experiences and narratives lead to judgemental bias. Subjective interpretation of situations can be criticised as not being sufficient for sound ethical deliberation. Subjective anticipations can enrich the ethical deliberation, but it is obvious that by making this decision Mrs. F. was so captured in her experience with a single case that she did not pay attention to the complex medical prognosis, and

---

[17] The case was presented in one of the detailed EDIG-interviews in Leuzinger-Bohleber et al. (Eds.) 2008b, pp. 188–195.

the considerable medical achievements which would have contributed to a very different quality of life for her child than her brother had decades ago. Her vision about the poor quality of life for her future child did not necessarily correspond to actual facts. Since then medicine has developed a safe and effective therapy for haemophilia so that life for people with this disorder is not comparable with her personal experiences. Experiences and narratives help pregnant women in making visions concrete, but the emphasis on these subjective experiences may be problematic for moral deliberation if they do not reflect the experiences of comparable others.

A third conflict between the two kinds of moral reasoning is linked to the previous one: emotions may cause psychological pressure to have PND even if there is no reasonable balance between the risk of harm to the fetus and the risk of having genetic disorders. The anxiety of having a disabled child on the one hand, and the trust in the technology offered, as well as the hope for medical control on the other may become so strong that pregnant women prefer to undergo PND even if it causes a higher risk than their base line risk of having a child with a disability. Decisions based on emotions like anxiety are very authentic but they tend to exclude the rational balancing of probabilities and to disregard severity, the kinds of symptoms and the time issue. Professional guidelines and law makers introduced limitations in the offer of PND based on this rational balance (e.g. amniocentesis only above 35 or 39 years) even if women would prefer to have more security through technology (Boyd et al. 2008). In my view these policies are not underestimating the emotional burden of uncertainty of pregnant women but they see professional responsibility in defining rational balances between the risks at stake. Such limits are not always agreed to by pregnant women.

9.6 Conclusions

Summarizing the inherent ethical dilemma of PND, a key question is what are the benefits and harms of the test. Benefits and harms are different for the fetus and for the pregnant women. Since there are no prenatal therapies for the conditions tested for a genetic test brings, most of the time, no benefits for the fetus.[18] Furthermore, PND always has a small chance of considerable harm to the fetus. On the other hand, harms of PND for pregnant women are not negligible but can be considered as rather small. More difficult is to determine the benefits of PND for her. In rare cases prenatal diagnostics may detect anomalies (like ectopic pregnancy) that may have lethal consequences for the pregnant woman. In such a situation she has a definite benefit from the test. Less clear is the case if the detected disorder is lethal for the fetus. In this case the pregnant woman has some benefit, if she needs this information, in preparing herself for the time after delivery or if she wants to terminate the pregnancy. The most problematic situation is if the detected genetic

---

[18] Conditions amenable to prenatal treatment of the fetus have to be treated differently.

conditions are not lethal. In this case there are different legal regulations in Europe – but all countries agree that termination of a late pregnancy cannot be justified by the mere wish of pregnant women (Boyd et al. 2008). Justifications for late abortions of fetuses with non lethal anomalies are commonly associated with both a medical condition of the fetus and the psychological and social resources of the pregnant woman. If she is worried by visions of having a child with disability she may opt for the test in order to exclude the suspected conditions. This is initially of psychological benefit. Here I assumed that psychological benefits like the tranquillizing effect are as important in counselling as physiological benefits. Therefore if psychological benefits are considerable they may outweigh the smaller physiological risks to the fetus. The balance between the benefits and the possible and actual harms must, however, be stressed more than in cases of suspected lethal anomalies.

Pregnant women have a different perspective on the pregnancy than medical professionals and this perspective is necessary in order to individualise the decision. At the same time pregnant women have different worries about the health of the future child. They might wish "to do everything possible". PND can reduce these worries effectively if counselling mainly focuses on the benefits of the genetic test and the power of the available technology. Even if there is a demand, the pure argument of psychological relief seems not to justify the use of PND. Discussion about the possibility of finding an abnormality as well as the limitations of the prognostic power of PND must be part of the counselling, even if this is counterproductive concerning one important goal of the test, i.e. psychological relief from an unbearable uncertainty.

It is obvious that the two ethical concepts, the ethical reasoning of health care professionals and the ethical reasoning of pregnant women, are different and can lead to conflicts. Thus the two basic principles of decision-making are not to interpret in absolute sense: Professional non-directiveness is not the same as eliminating all kinds of professional normativity. And freedom of choice is not decision-making by respecting only subjectively relevant facts.

Neither paternalistic judgements of the care giver nor the subjective estimations of the pregnant women alone can serve as reliable ethical justification for complex decisions. Therefore, non-directive counselling means not only providing information and receiving the decision but presenting the norms of the profession, making sure that pregnant women understood all the information properly, challenging their perspective and balancing the two kinds of ethical reasoning. This balancing has to be done together by counsellors and pregnant women.

References

Boyd PA, DeVigan C, Khoshnood B, Loane M, Garne E, Dolk H, The EUROCAT working group (2008) Survey of prenatal screening policies in Europe for structural malformations and chromosome anomalies, and their impact on detection and termination rates for neural tube defects and Down's syndrome. BJOG An Int J Obstet Gynaecol 115:689–696

Bramwell R, West H, Salmon P (2006) Health professionals' and service users' interpretation of screening test results: experimental study. Br Med J 333:284–286

Bundeszentrale für gesundheitliche Aufklärung (BZgA) (2006) Schwangerschaftserleben und Pränataldiagnostik. Repräsentative Befragung Schwangerer zum Thema Pränataldiagnostik. BZgA, Köln

Corandi E (2001) Take care: Grundlagen einer Ethik der Achtsamkeit. Campus, Frankfurt

Cremer M, Liebe M, Schnobel R (1983) Genetische Beratung in Heidelberg von Januar 1979 bis einschließlich Juni 1982. Bericht an das Bundesministerium für Jugend, Familie und Gesundheit über das Modellprojekt: "Modell zur ausreichenden Versorgung der Bevölkerung einer Region mit genetischer Präventiv-Medizin in Kooperation zwischen einer genetischen Beratungsstelle und dem Öffentlichen Gesundheitsdienst. Entwicklung eines Satellitensystems für den Rhein-Neckar-Raum". Universität Heidelberg, Heidelberg

Czeizel E (1981) Genetikai tanácsadás. Elmélet és módszer. Medicina Könyvkiadó, Budapest

Decruyenaere M, Evers-Kiebooms G, Boogaerts A, Philippe K, Demyttenaere K, Dom R, Vandenberghe W, Fryns J-P (2007) The complexity of reproductive decision-making in asymptomatic carriers of the Huntington mutation. Eur J Hum Genet 15:453–462

Engels E-M (2008) Experience and ethics: ethical and methodological reflections on the integration of the EDIG study in the ethical landscape. In: Leuziger-Bohleber M, Engels E-M, Tsiantis J (eds.) The janus face of prenatal diagnostics. Karnac, London, pp 251–272

Fischmann T, Pfenning N, Läzer KL, Rüger B, Tzivoni Y, Vassilopoulou V, Ladopoulou K, Bianchi I, Fiandaca D, Sarchi F (2008) Empirical data evaluation on EDIG (Ethical dilemmas due to prenatal an genetic diagnostics). In: Leuziger-Bohleber M, Engels E-M, Tsiantis J (eds) The janus face of prenatal diagnostics. Karnac, London, pp 89–135

Gilligan C (1984) Die andere Stimme. Lebenskonflikte und die Moral der Frau. Pieper, München (orig. ed.: Gilligan C (1982) In a different voice. Harvard University Press, Cambridge)

International Huntington Association, IHA (2008) http://www.huntington-assoc.com/. Accessed 10 Oct 2008

Jungermann H, Franke G, Schneider B (1981) Beratung bei Schwangerschaftskonflikten. Bericht über die Entwicklung und Erprobung eines Modells zur sozialen Beratung gemäß § 218. Kohlhammer, Stuttgart

Kirmse B (2006) Duchenne muscular dystrophy. In: Medical encyclopedia. http://www.nlm.nih.gov/MEDLINEPLUS/ency/article/000705.htm. Accessed 10 Oct 2008

Koch T (2008) Is Tom Shakespeare disabled? J Med Ethics 34:18–20

Komaromy M (2000a) What is FAP? In: Genetic health. http://www.genetichealth.com/CRC_FAP_A_Hereditary_Syndrome.shtml. Accessed 10 Oct 2008

Komaromy M (2000b) What is HNPCC? In: Genetic health. http://www.genetichealth.com/CRC_HNPCC_A_Hereditary_Syndrome.shtml. Accessed 10 Oct 2008

Korenromp MJ, Page-Christiaens G, van den Bout J, Mulder E, Visser G (2007) Maternal decisions to terminate pregnancy in case of Down syndrome. Am J Obstet Gynecol 196:149.e1–149.e11

Kovács L (2008) Prädiktive genetische Beratung in Deutschland – Eine empirische Studie. Institute for Advanced Studies, Vienna

Kovács L, Frewer A (2009) Die Macht medizinischer Metaphern: Studien zur Bildersprache genetischer Beratung und ihren ethischen Implikationen. In: Hirschberg I, Grießler E, Littig B, Frewer A (eds) Ethische Fragen genetischer Beratung: Klinische Erfahrungen, Forschungsstudien und soziale Perspektiven. Peter Lang, Frankfurt, pp 205–221

Leuzinger-Bohleber M, Belz A, Caverzasi E, Fischmann T, Hau S, Tsiantis J, Tzavaras N (2008a) Interviewing women and couples after prenatal and genetic diagnostics. In: Leuziger-Bohleber M, Engels E-M, Tsiantis J (eds.) The janus face of prenatal diagnostics: a European study bridging ethics, psychoanalysis, and medicine. Karnac Books Ltd, London, pp 151–218

Leuzinger-Bohleber M, Engels E-M, Tsiantis J (eds.) (2008b) The janus face of prenatal diagnostics: a European study bridging ethics, psychoanalysis, and medicine. Karnac Books Ltd, London

Lubs HA, de la Cruz F (eds.) (1977) Genetic counselling. Raven, New York

Lux V (2005) Pränataldiagnostik in der Schwangerenvorsorge und der Schwangerschaftsabbruch nach Pränataldiagnostik. IMEW, Berlin

Mahn H (1979) Wirkung der genetischen Beratung. In: Bundesministerium für Jugend, Familie und Gesundheit (ed.) Genetische Beratung. Ein Modellversuch der Bundesregierung in Frankfurt und Marburg. Verlag Bundesministerium für Jugend, Familie und Gesundheit, Bad Godesberg, pp 86–96

Malenka DJ, Baron JA, Johansen S, Wahrenberger JW, Ross JM (1993) The framing effect of relative and absolute risk. J Gen Intern Med 8:543–548

Porter IH (1977) Evolution of genetic counselling in America. In: Lubs HA, de la Cruz F (eds.) Genetic counseling. Raven, New York, pp 17–31

Rogers CR (1942) Counseling and psychotherapy. Houghton Miflin Comp, Boston

Rothschild J (2005) The dream of the perfect child. Indiana University Press, Bloomington

Slovic P, Fischhoff B, Lichtenstein S (1982) Facts versus fears: understanding perceived risk. In: Kahnemann D, Slovic P, Tversky A (eds.) Judgement under uncertainty: heuristics and biases. Cambridge University Press, Cambridge, pp 463–489

Tversky A, Kahneman D (1973) Availability: a heuristic for judging frequency and probability. Cogn Psychol 5:207–232

Wieser B (2006) Translating medical practices: an action-network theory perspective. In: Wieser B, Karner S, Berger W (eds.) Prenatal testing: individual decision or distributed action? Profil, München, Wien, pp 101–129

Wüstner K (2001) Subjektive Wahrscheinlichkeiten in der genetischen Beratung. Z Gesundheitswissenschaften 9:8–23

Chapter 10
Ethical Thoughts on Counselling and Accompanying Women and Couples Before, During and After Prenatal Diagnosis

Dierk Starnitzke

Abstract This chapter attempts to reflect on the clearly increasing types of problems in the context of counselling in prenatal diagnosis (PND) from an ethical perspective. As a result of my experience in my present function, as speaker of the board of one of the largest institutions for disabled people, it is important for me to point out the following: a fetus with a disability, after a positive diagnosis from PND, is in general terminated but can, in my opinion, live his own life after his birth if someone provides appropriate support. The life of this person may, depending on the type of disability, last either only a few hours or many years. Therefore counselling during PND should keep the decision of the termination or continuation of a pregnancy open. To ensure this, you need access to a wide range of counselling, in an interdisciplinary setting, including at least medical, psychosocial and ethical aspects.

Keywords Genetic counselling • Prenatal diagnosis • Psychosocial aspects

10.1 Introduction

My thoughts are influenced by three positions:

First, during my years as head of a large institution providing integrated support for people with disabilities, who are nevertheless able to live a good life, I gained experiences which might be relevant to people with a positive prenatal diagnosis of a particular illness or disability. The Diakonische Stiftung Wittekindshof, of which I am Spokesman of the Board, has over 2.800 employees in Nordrhein-Westfalen, and supports around 3.400 people, some with very extreme disabilities.

Secondly, I had the opportunity to participate in the training for Psychosocial Counselling of Prenatal Diagnostics at the Evangelische Zentralinstitut of the EKD for family counselling (EZI) in Berlin. This was a pilot project supported by the

---

D. Starnitzke (✉)
Diakonische Stiftung Wittekindshof, Zur Kirche 2, Bad Oeynhausen 32549, Germany
e-mail: dierk.starnitzke@wittekindshof.de

responsible ministry. The work with the counsellors and the debates about their examples from practical experiences were very moving for me and led me to think about current processes in society and medical ethics.

Thirdly, my experience in both of these areas leads me, as a scientific theologian, teaching in the Church University Wuppertal/Bethel, to reflect theologically on those questions. Bethel is also a place where many people live with serious disabilities and illnesses.

10.2 The Current Development of Prenatal Diagnostics

PND is a medical field where one can expect that, in the near future the management of pregnancies, at least in Germany, will change quite dramatically. In future even during first time routine examinations, abnormalities of the fetus will become increasingly visible, at a time when the pregnant woman has not yet concerned herself with the subsequent medical and ethical questions. In the past, the early diagnosis of fetal anomalies was quite rare. With the increased use of an improvement in ultrasound equipment, genetic analysis and other diagnostic methods, a larger proportion of pregnant women will be confronted with these questions. In the event of a diagnosis of fetal anomaly, even in the early stages of pregnancy, decisions have to be made by the concerned parties and in particular by the pregnant woman. Questions raised by this are: should further medical diagnostic procedures be used under any circumstance to identify an abnormality? If not, how should one deal with the uncertainty of not knowing the exact diagnosis and which of the therapeutic possibilities will the expected child be deprived of because of this decision? If yes, up to what point should systematic medical diagnosis be carried out?

These are decidedly sensitive and difficult questions. Psychosocial counselling before, during and after PND has the task of supporting the mother or parents in coming to terms with their situation. I believe that psychosocial counselling in these cases reveals new ethical problems, which were not even considered in ethical debates before the start of PND in its present and future forms. Therefore, extensive ethical approaches need to be developed which take into account the complexity of the subject. There is, however, in general already a certain system – based on past experiences of pregnancy conflicts without PND – which deals with ethical questions of abortion. However, the ethical problems become more intense especially where PND is involved – where something becomes visible which formerly has been hidden, namely the developing fetus and its possible illnesses and/or disabilities. A comparable area of ethical questions can perhaps be found in pre-implantation diagnostics (PID), which is currently, and will continue to be performed. Compared to PID the questions concerning PND are far more existentially important because the development of the fetus is much further advanced, and moreover the attachment between the fetus and the pregnant woman is much more advanced.

10.3 Ethical Questions on PND in Context of Today's Society

One could first of all assume with regards to the difficult ethical questions which arise during the counselling about PND, that it could be useful to possess a specific set of values and take heed of them. Living in a clearly defined religious context, for instance, greatly influences how one deals with a diagnosis of disability and illness. The decision for or against a termination of pregnancy is influenced by this context as is how one handles the consequences of the decision, such as how to dispose of a dead fetus or how to support and accompany the child after birth as much as is possible throughout his life.

The current situation in German society is predominantly governed by the fact that there is a plurality of philosophies of life. During the training for PND in the EZI in Berlin it became clear that the consultants as well as the consultees came from various philosophical, religious and ideological backgrounds. Accordingly, the related values of orientation also differed, making counselling somewhat ambiguous.

None of the various religions and philosophies of life can, at any rate in our country, claim to offer the one and only appropriate way to handle the described questions. Therefore, one can hardly expect a general social consensus relating to abortions as a consequence of PND in such a pluralistic society. But, one can demand that at least a counselling institution or a training centre for counselling, like for example the EZI, should develop a consensus for its own work, which could then be implemented within its own institution. How this can be achieved, I will specify at the end of this article.

Considering the impossibility of formulating a general consensus of values in an ideologically plural society, it is my impression that a social consensus already exists on such ethical questions. However, I believe, that this social consensus is not based on medical, political, ethical or religious professional discussions, but rather a result of direct everyday practice in the context of PND. Decisions made by the concerned parties are not based on particular ethical principles but rather a particular social practice is developed in response to these questions, which then creates a certain ethos, i.e. a generally accepted attitude.

This is disturbing, as one would expect a higher variability of practice based on the plurality of philosophies of life in Germany and professional ethical research should pay great attention to this common, but ethically quite unreflected practice.

From my experience and perception there are, in the meantime, clear expectations that pregnant women with a positive diagnosis – that means where a disability or illness is ascertained or can be assumed with a high probability – should terminate the pregnancy. But in principle they are of course free to make their own decision. Nevertheless, in most situations in which such diagnoses occur, the readiness of the concerned parties to accept illnesses or disabilities is usually relatively small. In most cases therefore an early termination of the pregnancy and the death of the fetus is to be expected. For instance only a small percentage of fetuses which are diagnosed with Down syndrome (Trisomy 21) are accepted and born. But also in cases of more moderate types of physical and intellectual disabilities we are seeing a surprising readiness to end the pregnancy. In her most remarkable work Christiane Kohler-Weiß (2003) labeled this selection:

> Während die persönliche Entscheidung für einen Schwangerschaftsabbruch nach PND in individual-ethischer Perspektive gute Gründe haben kann, ist die PND, seit sie routinemäßig angewandt wird, in sozialethischer Perspektive als Selektionsinstrument zu bewerten. (Whilst personal decisions about the termination of a pregnancy after PND can have good reasons in an individual ethical perspective, PND is to be regarded since its routine implementation, from a social ethical point of view, as a selection instrument. Kohler-Weiß 2003, p. 410, translation D.S.)

This is not about prematurely judging these tendencies, morally or ethically. Adequate ethical recognition must be given to common social practices with respect to the questions referred to, before they lead to ethical statements, not least because most of the terminations of pregnancy have serious personal reasons.

Nevertheless, I find it helpful, that apart from the individual point of view, Kohler-Weiß also speaks about a fundamental social ethical viewpoint. The problems referred to applied in the past to limited numbers of pregnancies where such diagnostic techniques were used. In the future, earlier and more widespread possibilities will be available, for example through further improvements in ultrasound equipment and other diagnostic methods to indicate illnesses and disabilities, perhaps even with the first routine test. Then the medical system takes on a life of its own and women might feel pressured to go on with other tests. As a result, it will become more difficult for pregnant women who, for individual reasons, decide to withdraw from further diagnostics. It is realistic to expect that in the future a large proportion of prenatal disabilities and illnesses will be identified at a very early stage and that this will, in the majority of cases, lead to a "premature termination of the pregnancy". (I was asked by the chief physician of the gynaecological department of a protestant hospital, not to refer to this as the interruption of the pregnancy or even an abortion, but rather a 'premature termination'). One could assume that on the basis of such a developing social ethos, decisions made by individuals against this ethos would constitute an exception. In other words, ethical questions should not only be considered from the standpoint of individual ethics but also from the one of social ethics and individual decisions against the current social ethos should still be possible and supported.

One more essential is, that if in the future considerably fewer children with disabilities and illnesses are born, the possibility to gain positive experiences of meeting people with such illnesses and disabilities will be considerably reduced. This lessening in experience then reduces the readiness to accept a fetus with such disabilities and illnesses, et cetera. As an academic theologian and head of one of the largest institutions for disabled people, I would like to explicitly point out that it can be a great enrichment to life, having contact with such ill and disabled people. In order for society to enjoy such enriching experiences, an ethos is required which accepts the possibility of the birth and life of an ill or disabled person, at least as much as the possibility of termination.

10.4 Impressions on Current Counselling Practice During PND

The advantage of, and not to be missed opportunity of, interdisciplinary debates and research projects like the EDIG study on ethical dilemmas due to prenatal and genetic diagnostics is, in my opinion, that the counsellors can clearly and openly exchange their experiences concerning developing practice alongside academic discussions about ethical principles. These discussions need of course to be carried out with all seriousness and to a very high standard. They may be extraordinarily enriching, with respect to ethical academic questions, especially when confronted with the everyday counselling experiences of the course participants of the pilot project in the EZI. It became clear to me that problem situations as described by this group need to be addressed more in ethical, theological and also political discussions.

I would therefore like to raise two points, which became particularly clear to me during my work with the counsellors and with the case examples presented there, which I consider important for an appropriate academic discussion about the related issues.

Firstly, it is obvious that counselling is incorporated into systemic processes: as soon as the psychosocial counselling begins, the medical system has already begun to work in parallel. In doing so the medical system operates entirely within its own logic, which is orientated to the code sick/healthy (cf. Luhmann 1990). The processes within this system are very clearly structured; probabilities are raised for the existence of a particular illness or disability and thus the expected possibility of life or limitations of life of the fetus. Therefore medical diagnosis seems to be somewhat isolated from the social and psychological context of which it is actually a part. As a result, the pregnant woman and her partner, might feel unnoticed or unacknowledged as individuals. The medical system processes operate seemingly separate from areas such as the psychosocial counselling process. Therefore, it appears extremely difficult for doctors to engage with psychosocial counselling and it is not my intention to hastily and ethically disqualify these medical system processes. The medical system, however, owes its high performance ability to the fact that it is able to close itself off from the influence of other systems and act strongly on its own code – sick/healthy. (Concerning such systemic processes cf. Starnitzke 1996, p. 248 et seq.)

Psychosocial counselling, therefore, from a realistic point of view will have to establish its own system alongside the medical system. It will however be important to develop mutual relationships and transparency especially through closer cooperation and communication between doctors and psychosocial counsellors. However, these two models will continue to differ.

Secondly, a realistic ethical assessment of the counselling process must keep in mind the limitations of time within this process. In the event of a positive medical diagnosis, some serious decisions must be made in the following days and weeks about any further actions. As a result, time and the possibility to provide adequate information and counselling is extremely limited. In general, they cannot seriously incorporate ethical or psychosocial aspects, whilst also needing to take into account the extremely limited time-scale.

10.5 The Necessity of a Detailed Individual Ethical Reflective Consultation

Many difficult questions arise in association with positive diagnoses. This includes understanding the diagnosis, getting further information and making the decision to have either an abortion or to give birth to the affected fetus. The associated moral conflicts and being able to endure and cope with the abortion and separation from a fetus growing in one's own body, demand, more than anything: plenty of time! In professional counselling sessions such questions can be addressed by psychosocial counselling much better than by the medical system. But here also, the few sessions which do take place are considerably time limited. Within these constraints, extensive effort is made to enable the pregnant woman to find out more about or consider her medical, financial, social and psychological situation – albeit in a limited way. For a realistic consideration of the various options, including ethical viewpoints or even the initiation and accompaniment of a deep decision process, time is too short.

During the course in the EZI it became clear that, in a large proportion of the case reports discussed, the individuals being counselled came mainly from quite simple surroundings with a relatively low education level. They were not familiar with their medical circumstances and for this reason had difficulties in perceiving their own situation in an adequate way. Apart from this, they also knew very little of the possible support available after the birth of a disabled child (cf. EZI 2006, p. 239, Fig. 12). It was a challenging task for the consultants to explain to the couples difficult medical issues and possibilities. In current practice, to then consider the resulting problems on an ethical level, and assist the people concerned in reaching a balanced and ethically reflected decision, is in most cases almost impossible. It would demand a larger time-frame for intellectual and ethical reflection.

We must also bear in mind that counselling is also, for the consultants, a communicative, psychic and intellectually difficult task, which regularly involves them in excessively demanding situations. Consultants confront far reaching fundamental questions about life, which they also then have to come to terms with. At the same time, it is difficult for the consultants to disregard their own attitudes to life and not to influence the consultees. In addition, there are the external conditions – like the time constraints, which are challenging not only for the consultees but also for the consultants. Understandably, as a result, the consultants are taken to the limits of their capabilities and need supervision for themselves. Therefore, it would be desirable to introduce inter-disciplinary guidance groups where it would be possible to exchange personal and professional counselling experiences. These groups should at least consist of doctors, psychosocial consultants and ethical specialists, for example pastors.

In addition, I would like to mention the following: dealing with guilt, supporting parents during the termination of pregnancy, funeral rites and rituals etc., all these issues cannot be accomplished within the present existing counselling settings. These issues arise, nevertheless, and need to be dealt with by all involved. You have in any case not only physical, but also deep mental injuries of the couples and it is necessary to treat them, sooner or later. The best time would be in my opinion close to the injury.

For this reason, in my view, it is absolutely essential to organise a network of interdisciplinary counselling, which could at least begin to answer these questions. This would also involve further extensive training of the counsellors and supervising them in their counselling procedure and the creation of an overall framework within which the problems described here can be adequately dealt with.

10.6 Conclusion

On the basis of my comments, in closing I would like to formulate the following theses for further consideration:

1. With respect to the extraordinary high requirements, which are present in psychosocial PND counselling, I think it is necessary to expand the framework of counselling considerably. An appropriate consultation for PND requires, amongst others, extensive consideration of the related ethical questions. As I have tried to express here, they are so complex that a time frame should be made available for counselling, that is as long as the couples require.
2. A fundamental characteristic of a reasonable counselling session under the present day conditions should be that the consultation has an unbiased outcome. It could, instead of a relatively one sided arrangement to end the pregnancy, as it appears to be in the present general society, make an effort to have an open counselling process in which the continuation of the pregnancy is raised as a serious possibility in each case.
3. If, within the framework of counselling, we really want to make it possible to discuss these ethical questions, then the scale of the counselling service needs to be considerably expanded. On the one hand a limited counselling on late abortions is compulsory. But on the other hand you also need ethical specialists, who can add to this counselling setting. Therefore, it is necessary to improve interdisciplinary counselling networks for example of doctors, psychosocial and ethical professions.
4. Beyond the individual counselling, society as a whole should also be urged to engage with the ethical issues raised by PND. The extension of an interdisciplinary counselling process could be seen from a societal point of view as having an important social ethical effect. For every accordingly counselled pregnancy, there would be at least the possibility to reflect upon the reasons for continuing or ending a pregnancy more carefully. This would contribute to a widespread ethical reflection process within society.

5. A few medical centres have introduced such networks of interdisciplinary counselling, such as the prenatal centre am Kuhdamm in Berlin, for women who are more than 20 weeks pregnant. To accomplish its task there needs to be sufficient time to address difficult questions in the case of individual conflicts. Everything which could not be clarified through such counselling, must then in any case be painstakingly dealt with later by others, for example psychoanalysts and pastors – or, to the distress of the parties involved, remain not dealt with and accordingly a burden.

References

EZI, Evangelisches Zentralinstitut für Familienberatung GmbH (Eds.) (2006) Abschlussbericht zum Modellprojekt. Entwicklung, Erprobung und Evaluation eines Curriculums für die Beratung im Zusammenhang mit vorgeburtlichen Untersuchungen (Pränataldiagnostik) und bei zu erwartender Behinderung des Kindes, Berlin

Kohler-Weiß C (2003) Schutz der Menschwerdung. Schwangerschaft und Schwangerschaftskonflikt als Themen evangelischer Ethik. Gütersloher Verlagshaus, Gütersloh

Luhmann N (1990) Der medizinische Code. In: Luhmann N (ed) Soziologische Aufklärung 5: Konstruktivistische Perspektiven. Westdeutscher Verlag, Opladen, pp 183–195

Starnitzke D (1996) Diakonie als soziales System: eine theologische Grundlegung diakonischer Praxis in Auseinandersetzung mit Niklas Luhmann. Verlag Kohlhammer, Stuttgart Translation assisted by Alan Scott

Chapter 11
Client, Patient, Subject; Whom Should We Treat? On the Significance of the Unconscious in Medical Care and Counselling

Yair Tzivoni

Abstract The significance of relating to people in the prenatal diagnostics decision-making process as *subjects* with an unconscious is studied in this article. It is done in comparison to treating them as *clients* or to treating them as *patients*. These three different ways of relating to the person in the decision making process – Client, Patient and Subject – are studied from an ethical point of view. It is argued that the consideration of the unconscious in this decision-making process could promote a more ethical and responsible way of working through this process for the person who has to decide and for the doctor/counsellor.

Keywords Counselling • Decision-making • Ethics • Prenatal diagnostics • Psychoanalysis

The predicament of having to decide upon the life or death of a fetus with a disorder has been carefully studied within the EDIG project. This discussion included the topic of the moral status of the fetus, and specifically – the question of the beginning of life or the beginning of the fetus' humanity. In this context I would like to consider the "moral status" of the people involved in the process of decision-making: the doctor or counsellor and the parent. I wish to discuss their "moral status" within their relationship.

Here are three options to define the parent in relation to the doctor:

1. Client
2. Patient
3. Subject

Y. Tzivoni (✉)
Department of Psychology, The Hebrew University of Jerusalem
and
Department B, Beer Yaacov Mental Health Center, 74 a Bnei Dan St., Tel Aviv, 62500, Israel
e-mail: ytzivoni@gmail.com

And here are related categories to define the doctor in relation to the parent:

1. Service provider
2. Knowledge provider
3. Another subject

The difference between these categories is not merely an issue for philosophers but rather an issue with grave implications for the professional ethics of caretakers, be it doctors, psychologists, genetic counsellors and others involved.

11.1 The Parent in Relation to the Doctor

The first two categories are relatively easy to explain:

Client: *The Compact Oxford English Dictionary* (Soanes and Hawker 2005) defines a client as *a person using the services of a professional person or organisation*. We tend to use this word in proximity with consumer or customer, a word that implies payment involved. In this context the doctor is the service provider and as we know when service is on offer: "the customer is always right". In the framework of decision-making, the service provider is expected to help the client fulfil her/his wishes. Nowadays, in many contexts it is considered more appropriate to call the people who you treat *clients* and not *patients* as this implies more autonomy and respect.

Patient: According to the *Compact Oxford English Dictionary* (Soanes and Hawker 2005) a patient is: *a person receiving or registered to receive medical treatment*. This is the usual way to talk about people seeking medical treatment, and within the decision-making process the patient comes to the doctor for advice, and seeks the doctor's knowledge.

But what kind of knowledge is required to decide on moral matters? What kind of knowledge does a person need, to decide if a fetus with a genetic disorder should live or not? As an example of the difficulties in helping someone make a decision, we can use a story told by Sartre in *Existentialism is a humanism* (Sartre 2007), where he tells of a student of his who wanted to go and fight in a war in which he believed, but who was afraid to leave his sick mother behind. He presented his dilemma to Sartre, explaining the emotions and ideas that drew him towards each option. Sartre tells us that one can choose one's adviser according to the answer that one wants. When the student came to him, he knew what kind of answer he would get and this is what he heard: you are free and you must choose, or in other words – invent. There is no general law of morality to guide you; there is no pillar of fire that guides us in this world. Could Sartre give another answer? Perhaps he should have calculated the relative contribution of a single soldier in a war compared with that of an only son to his mother. Sartre didn't think so, but in moral decisions related to prenatal diagnosis it is very tempting.

In our experience from the EDIG study we saw what kind of knowledge is provided to people who have to decide on the life of their fetus. In most cases people were given, or actively sought, a great deal of information: This could be statistics, information about the disorder and its implications, the chances of survival and the quality of life. In Israel, much of pregnancy nowadays is about statistics – for example the chance of Down syndrome. This is calculated according to the mother's age, and then with every test new probabilities are continually produced. But what is the meaning of a 1 in 350 chance of having a child with Down syndrome? Similarly what does 1 in 80 or 1 in 8 mean? How is this knowledge relevant for the decision-making process? This question is not easy to answer. We can start by examining this on the factual level.

The same information provided to different decision makers will lead to diverse interpretations according to their personality and personal history, nationality, cultural identity, religion, social class, and so forth. Consequently, different interpretations are bound to entail different and even opposite courses of action. The decision to terminate a pregnancy due to a given chance of a certain disorder could be the "common sense" for one person in one culture just as for another "common sense" the same figures will guide her to continue the life of a fetus. So let us be careful about common sense as we can see that it is not a good compass for ethical questions.

We can also think of the relativity of ethical norms that changes with the zeitgeist – the "common sense" ethics of today will probably change considerably in 20 years. For example, eating foie gras is becoming unacceptable. It wasn't so just a few years ago. This point is almost self evident and far more dramatic examples are not hard to find.

It is clear that something other than the "objective" data is leading these moral decisions. Perhaps we may consider the idea that this data can even serve to hide the real motives for the actions taken. A person might believe that the statistics are determining his course of action, whereas other factors might be much more influential. This goes beyond the cultural differences that are more easily observed. We saw other factors in the interviews, where decisions of women interviewed were often related to what we regarded as the influence of personal history and unconscious drives.

This brings us to the third and more vague category in attempting to define both parent and doctor – namely the Subject. I shall try to tackle the vagueness by presenting the question in the simplest way: What is a person? Let us try to answer this using the term 'Subject'. Most psychoanalytic thinkers will agree on a few things that the subject *is not*. The Subject is not the *I*. It is not the I who is pro choice or against abortions. It is not the I who knows the motives for his actions and can explain it very well. It is not the Ich (German for I) or, in the English standard edition of the writings of Sigmund Freud, the Ego that is the subject.

An old story tells of Rabbi Hilel who was asked to teach the foreigner the whole Torah "in a nutshell". Rabbi Hilel replied: "That which is hateful to you, do not do to your fellow. That is the whole Torah; the rest is the explanation; go and learn".

I think that an appropriate answer for the same question about psychoanalysis "in a nutshell" would be: The I *is not* the person…the rest is the explanation: go and learn. I believe that this is the essence of the Freudian revolution, i.e. the de-centring of the subject, the conception that the I (ego) is not the landlord of his own home.

This idea doesn't suit and sometimes even opposes some of the values of our society. We are encouraged to view people as free agents and, as long as they are adults and not insane, to consider their choices as legitimate and as something for which they are responsible. This helps us clear our moral conscience and sense of responsibility as counsellors or doctors as we let the responsible adults choose for themselves after we have presented the "objective" data. This view is manifested in notions such as "informed consent" and "non directive counselling". Supposedly, we are providing the data/information and they are expected to decide. How convenient for us. How potentially dangerous for society. Do we need new proof in the twenty-first century to realize that people are not innately rational and ethical creatures? Is it not clear that moral statements of people are profoundly influenced by internal drives and interests and social influences?

Things become even more complicated when taking into consideration the fact that the counsellor/doctor is also a subject, another being whose moral judgment is also influenced by the same factors. The doctor may have lots of knowledge about biology and statistics, but does he really know any better about the value of life of an unborn child? He may well know the meaning of a certain disorder and many other facts, but all this knowledge might mask the fact that his moral stance is driven by many things of which he is unconscious.

11.2 Some Reflections on Counselling

Another factor is the pressure that the doctor is under during the counselling process in prenatal diagnosis. In the interviews conducted in Israel we saw that people often wished for a doctor or counsellor who dares to speak, dares to be authoritative and recommend a course of action. This was usually the case when people somehow knew what they wanted, but preferred to feel that it was not their choice but that of another entity – preferably a so called "objective" medical authority who represents the "right thing to do". This dependency on the doctor/counsellor helps reduce the burden of the moral dilemma and therefore the person feels better about the decision being made.

In the interviews, we could also see signs of the rage that is evoked when doctors articulated views that were different from what people wanted to hear. In those cases, people consequently felt the doctor was arrogant and out of line. Can doctors and counsellors express their own views or at least not support the views of the parent? Do they have the authority to do it? Are they obliged to do it?

At this point both psychoanalysis and ethics can contribute. It is understanable that the person in deliberation usually wishes to reduce his dilemma, and to make up his mind so as to feel that he had done the "right thing". But is this our goal? Both psychoanalysis and ethics agree that reducing moral dilemma or in the wider context: "reducing suffering" is not the chief endeavour. In that respect, psychoanalysts and ethicists agree. Psychology and medicine in the same situation are more inclined to put the reduction of both suffering and of moral dilemma as a primary goal.

Interestingly, in the small print describing the work within the EDIG project I found these objectives: "Based on the results of part A and B of the study there will be a search for strategies that serve to reduce ethical dilemmas for individual couples in the context of prenatal genetic diagnosis." I believe that these lines express the instinctive and common approach of caregivers who focus on reducing suffering and strive to do so. What is more natural for a doctor than to relieve the pain of his patient? This can be referred to as the psychological-medical approach.

The aim of reducing suffering is not alien to psychoanalysis and I don't suppose that an ethicist will object to it either. Nevertheless, as we argued elsewhere (Benziman and Tzivoni 2008), this goal cannot stand alone, and it is not the primary goal of the ethicist or the psychoanalyst. A person may do terrible things without consciously experiencing a moral dilemma. A person may express no pain, but from a psychoanalytic view we will detect symptoms and acts that reveal limitation and difficulties in his life. From both perspectives, being human is much more than living without pain or enjoying ourselves as much as possible. The Greek aphorism "Know thyself" is still relevant. I think that it hints at the dignity that exists in self-knowledge – that a person grows and progresses through seeking and assimilating knowledge about himself. "Know thyself" wouldn't have been such an important command unless people didn't know a lot about themselves: about the forces that drive them and about what makes them say the things they say. The implication is that to know more about oneself is a virtue. If the reduction of mental suffering is our only aim, we risk losing our ethics and neglecting an important part of the humanity of ourselves and of the people we treat.

There are factors other than the ethics or virtue of self-knowledge that make it important to regard people as subjects, as subjects with (or of) an unconscious.

In *Civilization and its Discontents* Freud (1930), with an exceptional hindsight, warns us of the dangerous aggressive drives within human beings and of the dangers of group dynamics. This is relevant to our discussion since we (as part of the group) might often forget to ask "what is the reason for this action". We are in danger of accepting things at face value without trying to understand the hidden motives of various parties involved. Trying to remind ourselves that the unconscious is no less part of a person than his "I" might help us to listen to people in a different way – a way which allows us to attend to what is beyond what they literally say and what they consciously know about themselves.

That endeavour becomes more complicated in a world of consumers (clients) who are "always right". It is not easy to ask questions regarding the motive for an action taken by a person who pays. We could also notice the tension that can arise in this regard within the EDIG group internal discussions. Some members suggested possible interpretations of unconscious motives of people that were interviewed and those interpretations ignited conflict within the group as other members related to them as intrusive and even degrading. It does sound degrading to explain, in ways of which they would certainly not approve, the actions of people who contributed so much to make this research project happen. It is much more comfortable to relate to the altruistic and kind parts of the people in front of us then to their aggression and liabilities.

The thing is that in a very basic way, the unconscious is degrading … it can be an unpleasant experience to find out how much of what we preach and believe in is provoked by things less noble then the convincing words that we produce to justify our intentions. However, there is also liberation in this, as these "truths" that we hold often limit our lives and our openness to the world.

In a brilliant documentary called "The Century of Self," the director, Adam Curtis shows how psychoanalytic ideas were used by corporations and politicians to shape the right person-customer for them. What seems to be a triumph of democracy in the form of self-fulfilment and personal freedom is said to mask forces and interests that benefit from this self-centred culture. The documentary demonstrates how people's values and needs can be manipulated and how dangerous this can be.

This raises the question of personal freedom. Are we ever free? For the ongoing purposes of society and specifically capitalism we assume that we are. Allegedly, we choose what we want to buy and what we want to be. The word "subject" implies otherwise. We are always subjects of something. In Lacan's phrasing – we are the subjects of the unconscious. This is the hidden "client" in psychoanalysis and not the I of the person in front of us. Remembering this type of "client" in other fields could serve as a compass. Remembering that we, and the people who deal with moral dilemmas, have an unconscious could help us be more careful in this unmapped territory.

This is a sophisticated compass that doesn't show us directly where to go, but can always remind us to check for directions and ask more questions about where we are heading. The ethics of psychoanalysis that can guide us here are to be non-directive, but in a unique way. This doesn't necessarily mean that we must give lots of information to the person, and perhaps on the way imply our hidden agenda. Furthermore, it does not mean to confirm or deny the person's moral act, or to be the one who knows what is right and what is wrong. It is rather to encourage questioning of oneself, both on a personal level and as a group. It is sure to be a very difficult thing to carry out, since we cannot force self-knowledge on anybody, but we are entitled (and perhaps even obliged) to think independently of our patients. We should also keep in mind that "the client is always right" is a marketing strategy slogan and not an ethical rule of thumb.

Psychoanalysis helps expand moral responsibility as it doesn't accept the phrase "I didn't mean to…" A subject in psychoanalytic treatment learns that his "slips of the tongue" and "accidental mistakes" are no less his own than his well articulated words (perhaps even more so).

Let us now return to the doctor/counsellor facing the person who has to decide. In an environment deeply influenced by post modernistic ideas all could be concluded in a simplistic manner by saying that we all have our relative views of ethics and not one of us should consider himself more knowledgeable than the other. From this point of view the (pseudo?) non-directive attitude is the only one possible.

But as I tried to show previously, this attitude could serve as a disguise, a way to overlook the massive forces that direct the choices that people have to make. People have to *act*. Termination of pregnancy is an act, and the act has ethical significance. But what is an ethical act? In Lacanian psychoanalysis a basic characteristic of an *act* is that it is possible to consider the initiator of the act as responsible for it (Evans 1996). This doesn't mean to be responsible in the forensic sense. Rather, it is to be responsible in terms of intentionality, which includes unconscious motives and desire as well as overt and declared motives. The subject is asked to take responsibility for his desire. Taking this kind of responsibility makes what he does an *act* in the ethical way. This holds true for the doctor/counsellor as well. He cannot and should not avoid the *act*. His act has to do with what he says and what he doesn't say. He has to take responsibility for what drives him, and perhaps as an ideal, an ethical counsellor should aim towards helping the subject face the decision at hand and to make it an *act*.

11.3 Conclusion

Going back to the Client-Patient-Subject dilemma within the current context: we cannot truly expect to treat the person in front of us strictly as a subject, and to rigorously aim at making him responsible for his desire. This would be too pretentious. However – let us not stay with merely the Client approach that keeps us out of the ethics arena. This leaves us with Patient, but let us try to make him a subjective one. A patient who has an unconscious mind that is not less him then his conscious mind. The same stands for us. This goal could help us take fewer things for granted, and be a little more suspicious of our "common sense", even if it might not always be simpler or more pleasant in the counselling room.

References

Benziman Y, Tzivoni Y (2008) The interchange between psychoanalysis and philosophy in the understanding of ethical decision. In: Leuzinger-Bohleber M, Engels E-M, Tsiantis J (eds.) The janus face of prenatal diagnostics. Karnac, London, pp 327–343

Evans D (1996) An introductory dictionary of lacanian psychoanalysis. Published by Brunner-Routledge (now an imprint of Taylor and Francis Books Ltd) ISBN: 0415135230 (paperback)

Freud S (1930) Civilization and its discontents. In: James S, et al. (eds.) The standard edition of the complete psychological works of Sigmund Freud. The Hogart Press and the Institute of Psychoanalysis, London (1927–1931): The future of an illusion, civilization and its discontents, and other works, vol 21, pp 57–146

Sartre JP (2007) Existentialism is a humanism. Yale University Press, New Haven

Soanes C, Hawker S (eds.) (2005) Compact Oxford English dictionary of current English, 3rd edn. Oxford University Press, London

Chapter 12
Decision to Know and Decision to Act

Regina Sommer

Abstract Findings in prenatal diagnostics indicating a severe disorder of the fetus make it necessary for the woman to decide whether to abort or to give birth to the child. The decision about what to do will be elucidated by the Aristotelian model of decision-making. One aim is to raise the awareness of this decision by means of the finding in the empirical data of the EDIG study, namely that some women are saying that they made no decision. According to Aristotle, actions committed under compulsion or through ignorance are not decided by the agent. It will be argued that the women's decisions meet none of these criteria and that the women are free to make a decision. Nevertheless their choice is limited as they only opt for an action that fits within the plan of their lives.

Keywords Aristotle • Decision-making • Nicomachean Ethics • Prenatal diagnostics

12.1 Introduction

Progress in reproductive medicine opens up new possibilities for interfering with pregnancy that we did not have before. The confirmation of pregnancy by the physician marks the starting point of antenatal care with further examinations throughout pregnancy. Medical care responds to the wishes of the future mother, as she can choose which tests she wants to have done and which ones she doesn't. She needs to make decisions about tests even if everything is fine with the baby. If any aberrant prenatal test result is found that indicates a severe disorder, the woman finds herself having to decide for or against termination of pregnancy.

In this paper, that decision will be explored by means of the Aristotelian model of decision-making. In fact, Aristotle is one of the first philosophers to shape the notion of "decision" (προαίρεσις, "prohairesis") as the technical term we know today. By developing the idea of a person's responsibility for the actions he or she

R. Sommer (✉)
Ethics in the Life Sciences, Faculty of Science, Eberhard Karls Universität Tübingen
e-mail: Regina@Wittelsberger.com

performs, Aristotle differs from those who believe in fate as in the Greek tragedies where a person acts because she is predestined and cannot escape destiny (Gauthier and Jolif 1959). The study *Ethical Dilemmas due to Prenatal and Genetic Diagnostics* (EDIG) had the chance to benefit from interdisciplinary dialogue and has offered the opportunity to apply ethical reasoning to empirical findings of decision-making as described by physicians and psychologists. Results of the study can be found in the book "The Janus Face of Prenatal Diagnostics" (Leuzinger-Bohleber et al. 2008). The lively discussions were especially beneficial because theoretical reflection is rarely explored in detail in the context of everyday realities. The purpose here is to focus upon a particular finding in the empirical data of the EDIG study, that is that some women said that they made no decision, but that their action for them was obvious.[1] The question whether the conspicuous test results made a decision necessary was reported by some women with "no". This answer was reported whether the action was termination or continuation of the pregnancy. Do these women – facing an event of great importance as it concerns human life – deny responsibility and 'choose' the action without deciding? If they did not make the decision they did not act voluntarily and therefore are absolved from responsibility.

Circumstances in which it is not possible for someone to make a choice are illustrated in the Aristotelian model of decision-making. We shall look into this model to see if it can be applied to the situation of pregnant women.

In order to examine the notion of decision-making and compare it to the decision a woman takes after any findings in prenatal diagnostics, Aristotle's analysis in the Nicomachean Ethics (NE), Book III, is taken as reference. Moreover, the philosophical discussion of decision and the voluntary is based on the work of Kuhn (1960) and Rapp (1995).

According to Aristotle, decision-making belongs to the voluntary. Actions can be voluntary, involuntary or mixed (having voluntary and involuntary moments). Mixed actions are "voluntary, though perhaps involuntary apart from circumstances – for no one would choose to do any such action in and for itself" (NE 1110a19). In consideration of the women's declaration that they did not decide, their act was involuntary. According to Aristotle, actions are involuntary when done under compulsion or through ignorance, so we will have to consider if one of these stipulations is met.

12.2 Actions Under Compulsion

12.2.1 Pure Compulsion

Is it possible that the women made no decision because they felt some pressure from their partner, the physician, society, religion or other, in a way that one alternative, abortion or childbearing, was out of the question? That pressure may obviously

---

[1] Questionnaires were elaborated by the members of the EDIG study and filled in by the women having taken part in the study in Germany, not published.

force the pregnant woman into continuation of a pregnancy with an unwanted child or into abortion, or, at least, may switch the balance for one or the other. However, this is not the compulsion Aristotle has in mind. "An act is compulsory when its origin is from without, being of such a nature that the agent, or person compelled, contributes nothing to it" (NE 1110a2–1110a3). This definition makes the margin of involuntary actions carried out under compulsion very small as illustrated by Aristotle: "for example, when a ship's captain is carried somewhere by stress of the weather" (l.c.). The woman with the decision to be made is not in the same position, at the mercy of external factors, as the ship's captain, so her action cannot be called involuntary through compulsion.

12.2.2 Compulsion by Force of Circumstances

The decision the woman has to reach under this criterion applies to actions "done through fear of a worse alternative, or for some noble object" (NE 1110a4). Circumstances may compel a person to act in a way she would not have acted in the absence of these circumstances, e.g. "when cargo is jettisoned in a storm; apart from circumstances, no one voluntarily throws away his property, but to save his own life and that of his shipmates any sane man would do so" (NE 1110a5–1110a6). The circumstance in question is a pregnancy with an affected child. The situation makes it necessary for the woman to choose termination or continuation of the pregnancy. In both cases she does not really decide for one option but only against the other. In the questionnaires of the EDIG study, we find this dilemma expressed through the statement: "There is no right decision." Women who are pregnant with a child with an anomaly who choose a termination of pregnancy may do so because of fear, e.g. about life with a severely disabled child, or the unendurable suffering of the child, or the death of their child after the birth. If they choose to carry the child to term this might be because they fear the procedure of abortion or a life of self-reproach after an abortion. The definitive action is in both cases realised out of the negative, and they did not make their choice affirmatively, as no one would choose abortion or to have a disabled child for itself. An explanation of the fact that the women say they have no choice might be that there is indeed a compulsion: they have to take a decision. Nevertheless, in the decision-making itself, they are free.

Mixed actions are "to be pronounced intrinsically involuntary but voluntary in the circumstances, and in preference to the alternative" (NE 1110b9). An action that in general was to be avoided becomes in this concrete situation the preferred action. One further condition for actions to be voluntary is that they are "deliberately chosen" and that "their origin lies in the agent" (l.c.). The origin of termination or continuation of pregnancy lies in the pregnant woman. A deliberate choice (NE 1110b9), as a further condition for an action to be voluntary, can only be assured through knowledge about the action we are deciding on. Ignorance here would make a decision impossible.

12.3 Actions Through Ignorance

Actions are involuntary not only under compulsion, but also when made through ignorance. We do indeed accept excuses like "I'm sorry, I did not know". For a pregnant woman choosing prenatal diagnostics, misunderstanding of the test results cannot be excluded. She would have an excuse for her action if her decision would have been different if she had the right understanding of the test results. The question as to whether she has all the information she needs to decide is more difficult to answer than to show that her action was not compulsory. However, the women in question seem not to see this problem. Their answer that they did not decide indicates the forced need for a decision, created through the information they had and not out of lack of information. Nevertheless, one reason to follow Aristotle here is that he calls ignorant and thus done without decision an action where the intention is missed so that it turns out different than intended.[2]

According to Aristotle we may err in several elements in regard of an action, namely about "the agent, the act, the thing [...], the instrument [...], the effect [...], and the manner" (NE 1111a2ff). Examples in the Nicomachean Ethics show that Aristotle claims actions to be involuntary when they do not have the intended outcome, for instance "they let it off when they only meant to show how it worked as the prisoner pleaded in the catapult case" (NE 1111a11). In the context of this Aristotelian understanding of ignorance in prenatal diagnostics the following possibilities may be of relevance.

- One case concerns testing procedures which unintentionally induce abortion. The fatal test makes any further decision obsolete; prenatal diagnostics did not reach their aim, i.e. to obtain information and, in the case of a conspicuous test result, enable a decision for or against the birth of a disabled baby. This is the same when there is a simple loss of the embryo. In both cases the pregnant woman makes no choice.
- Another action through ignorance would be the choice for a termination of pregnancy, but the child survives the procedure. Law requires that everything be done to keep this child alive. The woman then would have the child she wanted to avoid, and her decision has proved to be irrelevant.[3] Because of the possibility that the fetus can survive an abortion beyond week 20/22 of pregnancy, feticide is nowadays usually undertaken prior to abortion.

---

[2] This might seem strange to us as it is not the understanding of ignorance we use today. For us, the miss of the intention is due to the risks one should take into consideration in decision-making.

[3] In the questionnaires, we find one woman explaining this difficulty as her fetus has developed so far that it can already survive independently. She tells that a recommendation by professionals was to give this child, if it survives, up for adoption. In Germany, the "Oldenburger Baby", a fetus that survived abortion in the 25th week of pregnancy in 1997, gained notoriety. After the missed abortion, it didn't die but it was left alone for hours before it got any medical support. Nowadays the child lives in a foster family. For more information look at http://www.tim-lebt.de

- Childbearing also becomes an act taken through ignorance when, because of negative test results, a pregnancy is continued and only after birth abnormalities are discovered. Conversely, in the case of a positive screening test result, where the probability of a disorder is expressed in percentages, in many cases the baby will not be affected. To choose termination of pregnancy because of an incorrect interpretation of the positive screening test result becomes hence involuntary.

The mistakes of actions committed because of ignorance have, for our purpose, to be enlarged in order to include the cases in which not only the intention of the action is missed. In fact, the problem of ignorance in decision-making is also perceivable in the following cases.

- Ignorance in the decision-making is possible for the pregnant woman and seems to make the decision itself disappear: a woman expecting a child can apparently avoid the decision and any test result by not having any tests at all. If the child is then born with severe abnormalities the mother can say she had no option but to give birth to that child because she did not know of the problems. Aristotle can help to clarify the question of responsibility here: Ignorance can only count as an excuse for the actor not to be responsible for the action if the ignorance was not self-inflicted (NE 1113b32). Nowadays, a pregnant woman choosing not to have prenatal testing to avoid the difficult decision of abortion or having the child if an abnormality was found, intended to maintain this ignorance and hence would stay responsible for the action, i.e. birth of a child that might have an abnormality. The action of childbearing here is not made due to a decision through ignorance but is due to a decision for ignorance.
- A decision to have or not to have an affected child assumes knowledge about the disorder compared with normality and about what it means for the child to have this condition. If the pregnant woman is aware that there is some lack of knowledge about the implications of the disorder, this increases uncertainty in decision-making. In order to clarify and explain the medical facts and to communicate with the pregnant woman, genetic counselling is very important. The pregnant woman is exposed to other viewpoints and may be more susceptible to those views if she is uncertain about what she wants to do. A discussion about genetic counselling goes beyond the scope of this contribution; in any case, the role of the counsellor is essential in the woman's decision.
- To provide information to the pregnant woman or to keep it from her may produce uncertainty and put pressure on her to think over again something she thought to be settled, as shown by Korenromp et al. (2007): the doubt the woman had in her decision-making was directly correlated with the feeling of guilt 4 months after termination of pregnancy. This finding leads Korenromp et al. (2007) to pose the question whether informed consent should always be the underlying principle of counselling, or whether it might be better not to present other alternatives for a mother who had already made a decision. Quadrelli et al. (2007) describe this scenario: that women didn't want to have further counselling and explanations concerning Down syndrome because they thought they had all the relevant information and had already made up their minds. This position

should be respected by counsellors, but it could also be a self-deception and cause later psychological strains.
- As the decision affects the future, its consequences are not fully known at the moment of deciding. It is therefore a decision made with a high degree of ignorance. The pregnant woman can neither completely appraise what it means to live with a severely disabled child nor the implications of her child's disorder. Can the future life of the child become fulfilled or happy in a sense and will it suffer or not? In the EDIG study, quality of life for the baby was an important reason for the decision of the respondents both to continue or to terminate the pregnancy. In the case of Down syndrome, a prediction about the baby's quality of life is especially difficult as one cannot predict the degree of disability and how it will affect the child's quality of life. Other disabilities also carry the same uncertainty about the severity of the disorder. This problem cannot be solved, as the decision whether to give birth to or to abort this child has to be taken before the facts can be known.
- The value of the information provided by a positive test result not only depends on the knowledge the pregnant woman already has, but it depends also on the quality of the test. In practice, with the offer of prenatal screening tests, e.g. a non-invasive screening measure like the maternal serum alpha fetoprotein test for neural tube defects (AFP testing), uncertainty about the test result is a characteristic of the test. The test is performed between 16 and 18 weeks of pregnancy, the results are available within 2–7 days. The disadvantage of this testing is that atypically high AFP levels can result from other factors than fetal abnormality. Approximately 20 of every 1,000 women screened obtain abnormally high results, but 19 of 20 of these conspicuous findings are false positive (Lobel et al. 2005). Thus a positive test result would then worry most woman unnecessarily, as only 1 in 20 (5%) of those tests with a positive result will go on to show a child with a neural tube defect.

12.4 Aristotelian Subtleness: Not Voluntary Actions

Aristotle calls actions happening through ignorance 'involuntary' only if the actor regrets his deed in retrospect. Actions done through ignorance but without causing sorrow are called "not voluntary" acts (NE 1110b20). In order to judge whether the action was done voluntarily, the feelings the actor has in retrospect must be taken into account. But these feelings are not available at the moment of decision-making. How can one decide freely, if the voluntary nature of the decision is only revealed after the event?

An explanation for the evaluation of an act afterwards is provided when we look at the presentation of decision and the voluntary in the Nicomachean Ethics. Aristotle points out that choice "appears to be intimately connected with virtue, and to afford a surer test of character than do our actions" (NE 1111b5). Sensations that do accompany the action allow an inference on the virtuousness of the person making the action and thus play a role in appraising the decision. It is now possible to look at feelings the women have in hindsight to evaluate whether the choice was made involuntarily or only not-voluntarily in cases where the women were not sufficiently informed. This distinction allows us to describe acts that are in fact done with a lack of knowledge and thus do not meet the criteria of being a voluntary act. Nevertheless, the act in question does not cause necessarily sorrow and may even be endorsed by the agent.

We will take a look at the questionnaires given to the women after they have terminated pregnancy or given birth to their child in order to include in our considerations the answers given after abortion or birth of the affected embryo.[4] Do the women believe that their choice was appropriate? Some women say in the questionnaire that they are convinced that their decision was right and are comfortable with it; some remarked that there is no right decision; but no one thinks they have chosen the wrong alternative. The action was thus voluntary or, if there is no remorse, not voluntary in the Aristotelian sense, but this action cannot be called involuntary. Either the woman acts voluntarily, knowing all relevant circumstances for the action; or she acts through ignorance, but since she has no regret, not voluntarily. In the empirical data we found no woman who expressed regret about the act she had chosen; maybe they really had none or maybe they did not admit it. Actions done through ignorance are thus not necessarily involuntary. Is it conceivable for a pregnant woman to stay ignorant deliberately, in order to act anyhow and remain indifferent about her choice because of her ignorance after the act has been accomplished? This complex question is debated controversially in literature; it leads to the practical question of whether the counsellor should always communicate all possible information to the pregnant woman to ensure a well thought out decision (see above).

It has been argued above that the statement of those women who claimed that they made no decision cannot be supported within the Aristotelian model. Their choice is not an action under compulsion, as they choose it themselves and are the origin of the action, i.e. the one that gets the chain of any further actions going. Nevertheless it may be an involuntary act because ignorance and misunderstanding cannot be excluded due to the very specific information of the test result and uncertainty about the child's real condition. The information the pregnant woman gets is of utmost importance to substantiate a decision, but the value of the information is conditional upon the way it is communicated, the appraisal of the future estimated from the test results, and to the limits of the test itself. One woman expresses doubt about the diagnosis (trisomy 18) and therefore challenges the chosen termination of pregnancy. But as she answers the question whether she took the right decision in the affirmative, her doubt is about the diagnosis itself and not about her decision what to do with her fetus that has trisomy 18. As the women do not regret their act it is, according to the Aristotelian model, not an involuntary act.

---

[4] Aim of this chapter is to explain not-voluntary actions in the Aristotelian understanding. For this analysis, we only have as criterion an action "causing sorrow" or "not causing sorrow". The exact differentiation of the time period elapsed after the event as we need it for any significant analysis if we look closer to the problem is blanked out.

12.5 Decision

Until now, a decision has been defined out of the negative, i.e. as preceding an act that happens neither under compulsion nor through ignorance. The positive aspect of a decision, pointed out by showing that it is also a conscious choice of one alternative, will be elaborated below. As we have seen, the statement of the women saying that they took no decision cannot be reconciled with the Aristotelian model of decision-making. We will now look more closely at Aristotle in order to appraise the choice of the pregnant women confronted with a positive test result as a genuine decision.

Decision or choice belongs to the voluntary and comes about through deliberation: "Perhaps we may define it as voluntary action preceded by deliberation, since choice involves reasoning and some process of thought" (NE 1113b16–1113b17). When women say that there was no decision, one reason could be that there was no proper deliberation as this is a condition for a choice to take place. If these women already had a preconceived opinion, they would not be interested in more information about alternatives. Nevertheless, they remain responsible for the action because they aim for this ignorance. The decision was then made previously without deliberation of some of the facts. This is the case when the mother wants no testing at all in order to avoid the need for action. Hence, avoidance of that decision is a precursor of decision itself, namely the decision to give birth irrespective of the child's condition.

"Choice will be a deliberative desire of things in our power;" (NE 1113b11). Aristotle's emphasis on desire and power can help to clarify the problem of pregnant women after any positive test result in prenatal diagnostics: the circumstances of the women are involuntary and not in their power. In contrast to wishes, which may also tend for impossibilities, e.g. to choose to have this baby without the disability, the decision is only about possible things. The test result pushes the women into the need to make the decision since the question about what they should do is in their power. The women in the study were all well aware of the choices they could make and the wishes for the baby that were not satisfiable.

Information from the test results that the pregnant women obtain is no end in itself. A voluntary act requires the necessary information, namely an "agent who knows the particular circumstances in which he is acting" (NE 1111a21). The women are aware that they need to know about the particular circumstances of the action in order to carry it out voluntarily, and they are aware that they cannot trust the information completely. Nevertheless this uncertainty relates to the information itself and to the particular circumstances they are deciding on. Searching for information therefore can also be understood as a limitation. It is not necessary to inform someone about everything we know, not necessary to ask for everything and satisfy one's curiosity, but only about the things we need to know for making a decision and implementing an action. With information as an end in itself there is the risk of getting lost in the process without reaching an endpoint; the possible actions may be seen as a mind game about desiderata. This endless deliberation may enable a person to arrive at an opinion, but may not solve the problem itself, because the very choice is omitted (NE 1113a3).

When a woman decides for or against continuation of a pregnancy, the implementation of her choice is also to be planned. "Previous deliberation seems to be implied by the very term of prohairetón, which denotes something chosen before other things" (NE 1112a17). The term "chosen before" has to be understood as referring to time. A decision concerns the question of what to do next in order to carry out the intentions of the agent. Because "the last step in the analysis seems to be the first step in the execution of the design" (NE 1112b12–1112b13), the question is always about how to act now, which means what to do next to accomplish one's intentions. The intentions are e.g. to promote life without suffering, and healthiness is essential for this aim; but if the unborn baby is severely affected and will suffer, it becomes inevitable to terminate pregnancy to meet the claim "promote life without suffering". Termination of pregnancy is hence the first step in the execution of the design, a life without suffering. When a woman gets knowledge about some aberrant test result she has to decide about the next step in the chain of activities to accomplish her aims.

12.6 Discussion

How can it be that a pregnant woman makes a decision while knowing that there is more information available but refuses it? The decision against more information without eliminating the decision for the action can be justified with Aristotle's assistance. We decide about the information we want to have according to the action we stand for:

- As has been shown, the information is targeted on action.
- The attitude towards information gives a hint for understanding the statement of the women saying they took no decision. Decision-making is integrated into a larger context: the inaugural sentence of the Nicomachean Ethics assigns to every decision to strive for a good. This grants to decisions a place in our life conduct: "Every art and every investigation, and likewise every practical pursuit or undertaking [προαίρεσις] seems to aim at some good: hence it is said that the Good is that at which all things aim" (NE 1094a1–1094a2). The Aristotelian investigation of decision-making in Book III of the Nicomachean Ethics is made in context of the discussion of virtues: "Virtue then is a settled disposition of the mind as regards the choice of actions and feelings" (NE 1106b36). To achieve a good life, we need to be virtuous and choose the actions in accordance with virtue. We have to take decisions that are demanded through the circumstances we are confronted with in our life, e.g. being pregnant with a disabled child. The decision we have to take belongs to our personal quest for happiness, we have to put it in our whole life concept. The meaning of a decision has therefore always to be regarded as being part of someone's life, e.g. in the questionnaires we find the question whether the woman felt any pressure to make a particular decision strongly denied with the explanation that "the decision affects after all our life".

Because we live differently and have different views because of the experiences we have had, not all possibilities for actions come into consideration for everyone. Some are excluded from the outset as they are in opposition to our personal life concept. The decision here and now is integrated into life as a whole. Then there is no room for any choice, not for every possible decision, but the action is concerted to the life-conduct of the agent.

12.7 Conclusion

When the pregnant woman is not compelled to the action and when she has all relevant information for the action, according to Aristotle, she actually makes a choice.

The idea to put the decision into the "life-frame" of the actor gives us a persuasive explanation and the possibility to put Aristotelian ethics together with the answers given by the women in the context of decision-making: it has been proved that the women do make a choice (according to Aristotle) as they are firstly not acting under compulsion; secondly, the role of information (in accordance with actions committed through ignorance in the narrow Aristotelian sense and in a broader common sense) has been investigated and it has thirdly been stressed that the whole deliberation is targeted on action. This fact allows fourthly to integrate decision-making into the Aristotelian doctrine of virtue and this gives us lastly a hint that not the decision alone is important, but that we have to consider the life of these women as a whole.

Possible actions are chosen by the woman beforehand and regulate her interest in information. Therefore she channels the information from the outset to steer the decision-making in a way that fits with her life-concept. The woman's choice arose out of her life conduct and did not take into account all of the possible actions. The reason for some women to say that they made no choice might be that their decision is subordinated to and determined by their life script.

References

Aristotle (1926) The Nicomachean Ethics. With an English translation by H. Rackham. Loeb Classical Library, William Heinemann, New York
Gauthier RA, Jolif JY (1959) L'Ethique à Nicomaque: Introduction, traduction et commentaire. Publications Universitaires de Louvain, Louvain
Korenromp M et al (2007) Maternal decision to terminate pregnancy after a diagnosis of Down syndrome. Am J Obstet 196:149.e1–149.e11
Kuhn H (1960) Der Begriff der Prohairesis in der Nikomachischen Ethik. In: Gadamer HG (ed.) Festschrift für. Mohr-Siebeck, Tübingen, pp 123–140
Leuzinger-Bohleber M, Engels E-M, Tsiantis J (2008) The janus face of prenatal diagnostics. a European study bridging ethics, psychoanalysis, and medicine. Karnac, London
Lobel M, Dias L, Meyer B (2005) Distress associated with prenatal screening for fetal abnormality. J Behav Med 1:65–68

Quadrelli R et al (2007) Parental decisions to abort or continue a pregnancy following prenatal diagnosis of chromosomal abnormalities in a setting where termination of pregnancy is not legally available. Prenat Diagn 27:228–232

Rapp C (1995) Freiwilligkeit, Entscheidung und Verantwortlichkeit. In: Höffe O (ed.) Aristoteles, Die Nikomachische Ethik. Akademie-Verlag, Berlin, pp 109–133

Chapter 13
Moral Decision-Making, Narratives and Genetic Diagnostics

Göran Collste

Abstract Information based on prenatal diagnosis (PND) may cause distress and confront potential parents with difficult choices. Four interviews of mothers in this situation of choice, performed within the EDIG project, are the basis for this article. The interviews reveal how the mothers experience the situation and how their decisions are influenced by their life stories. This article focuses on the moral aspects of their decision-making. The contextual information gives us a better insight into how human beings experience these kinds of difficult moral decisions. The stories can be interpreted in different ways. In my analysis I distinguish between explanations; what causes the decision; understanding, i.e. to trace the meaningfulness of the reasoning and justification; what is the normative motivation for the decision?

Keywords Empirical ethics • Life stories • Narrative ethics • Prenatal genetic diagnosis

13.1 Introduction

Health care is going through a diagnostic revolution. Due to new diagnostic technologies the possibilities to detect genetic disorders increase. However, with increased knowledge there follows new decisions and new responsibilities. This is particularly true in the area of prenatal diagnosis (PND). Through different diagnostic techniques more and more genetic information about the fetus is collected. This information may reassure parents that there are no signs of genetic dysfunctions. However, the information may also cause distress and confront the potential parents with difficult choices: should information that the expected child is carrying a genetic disorder lead to termination of pregnancy? Is there any possibility of treating the detected disorder?

G. Collste (✉)
Centre for Applied Ethics, Linköping University, Linköping 58183, Sweden
e-mail: goran.collste@liu.se

The empirical basis of this article is four interviews with prospective mothers undergoing PND carried out within the EDIG project. Three of the interviews are published in chapter six of the book *The Janus Face of Prenatal Diagnostics* (Leuzinger-Bohleber et al. 2008) and one (Mrs D) was circulated within the EDIG project group. The aim of my article is to reflect ethically upon and analyse ethical aspects of their reasoning and decision making (see also Hildt 2008).

The interviews present the mothers' own reflections about why they have a prenatal genetic test and how they use and interpret it. In particular, I will focus on their reasoning concerning the decision to keep or to abort the fetus as a consequence of the test result. The stories can be interpreted in different ways. In my analysis I distinguish between explanations: what causes the decision; understanding, i.e. to trace the meaningfulness of the reasoning, and justification; what is the normative motivation for the decision?

13.2 Theoretical Considerations

I will investigate how we, from an ethical point of view, can identify and understand the moral dilemmas facing couples undergoing prenatal diagnosis. How do they experience the problem? Why do they decide in a certain way? How do they live with their decision and do they regret anything?

When relating empirical investigations of this kind to ethics one enters into two kinds of ethical inquiries; empirical ethics and narrative ethics. The interviews show how the moral dilemmas are interpreted by those who are most concerned, i.e. the parents; what reasoning is behind a decision; and whether a decision is regretted or not. In *empirical ethics* one uses research methods from social sciences (for example interviews) in order to examine ethical issues (Medicine, Health Care and Philosophy 2004, special issue).

How shall we then interpret the stories told by parents? What do they tell us about, for example, a meaningful life, how to handle moral dilemmas and the (moral) characterisation of human action? What do the stories say about what is a good life for a particular person and perhaps about what is a good human life *per se*? How are the values and decisions embedded in social and moral settings (including moral traditions)? How are the decisions rationalised and how are they justified? These are the questions posed in *narrative ethics*. The idea to perceive and understand a person's decisions and actions, as part of his or her life story, is developed by e.g. philosophers Alasdair MacIntyre and Charles Taylor. MacIntyre emphasises that our lives are constituted by our contexts and that particular actions derive their character and meaning as parts of a whole. He writes (MacIntyre 1981, p. 208):

> We identify a particular action only by invoking two kinds of context…We place the agent's intentions…in causal and temporal order with reference to their role in his or her history; and we also place them with reference to their role in history of the setting or settings to which they belong…Narrative history of a certain kind turns out to be the basic and essential genre for the characterization of human action.

Charles Taylor sees a person's life as a continuum. Taylor writes (1989, p. 47):

…we need an orientation of the good…this sense of the good has to be woven into my understanding of my life as an unfolding story.

13.3 Interviews

I will summarise four interviews with prospective mothers, Mrs A-D, carried out within the EDIG project. My intention is neither to give a detailed nor a psychoanalytic account. I will restrict the presentation to some necessary information that can serve as a basis for a discussion of the ethical aspects of the four cases.

The stories are constructed through the interviews. This means that an interview gives the interviewee an opportunity to reflect on her present condition. She is encouraged to see her life in a social and historical context and perhaps also to see her life story as rational and meaningful. The construction of her personal narrative might imply some rationalisation but this fact does not make the interviews less relevant, although it is important to be aware of this fact. The story told is the story the interviewee wants us to hear.

The four interviews are different in character. Only one reveals parts of the woman's background and life story, but they all reveal feelings, intuitions and moral reflections in a situation of difficult decision-making and distress.

13.3.1 Interview A

Mrs A became pregnant at the age of 32. The result of amniocentesis was positive and indicated that the child was carrying *haemophilia*. On the basis of this result, Mrs A decided to abort.

The interview with Mrs A gives some information which explains Mrs A's decision to abort the fetus after the positive test result. She has already had a few abortions for psychological and social reasons. Hence, abortion was for her not something completely foreign. Mrs A's decision also seems to be influenced by her view of her mother. Her mother was always blaming others for her ill fate. In contrast to her mother, Mrs A wants to solve her problems by her own efforts. Her husband supported her decision to abort, but it was Mrs A who took the decision.

A key to understanding Mrs A's decision is her relationship with her brother. She was brought up with an older brother with *haemophilia*. Her brother's illness seems to have affected her childhood. As a girl she lived in the shadow of her brother. She was also accused by her mother of "lacking empathy" and not caring about his suffering. Hence, Mrs A has very bad memories of her brother's illness and thus she wants her own children, if possible, to avoid her brother's fate.

Obviously, we cannot understand Mrs A's decision to abort without knowing her life story. According to her narrative, the ill brother has in important ways shaped her life. Her main reason for interrupting the pregnancy, on the basis of genetic information, is that she does not want her child to live a life like her brother's. Without this intimate knowledge about a life with *haemophilia* her decision might have been different.

The ill brother also plays an important role in Mrs A's justification of her decision to abort. When informed about her decision the brother was hurt. Mrs A thinks that for him her decision implied "you don't want a person like me to live – you kill him". However, she makes a distinction between him as a person and his illness. She says that she did not want her own child suffering so much as her brother had during his whole life. Thus, her choice was directed towards the illness, not towards her brother. Furthermore, a cousin accuses her of choosing babies like a supermarket. Mrs A is offended by this comment but she maintains that the reason for her decision was to avoid her child suffering like her brother had done.

What, then, are the normative reasons behind her decision to abort? It seems that for Mrs A the overriding reason is to avoid suffering, which in this particular case implies avoiding the birth of a child with *haemophilia*. She does not seem to experience the situation as a moral dilemma because she does not consider the alternative which is to keep the child after being informed of the positive genetic diagnosis.

13.3.2 Interview B

Mrs B's story is focused on the discovery of the fetus' genetic condition and on her own feelings surrounding amniocentesis and abortion. Ultrasound investigation indicated that the fetus was carrying trisomy 18 (Edwards' syndrome), which leads to mental and physiological disabilities and, in the majority of cases, the death of the baby in less than 6 months after birth. Amniocentesis confirmed the diagnosis. According to Mrs B it was the doctor who recommended abortion when trisomy 18 was diagnosed. He said that with this information there was only one choice, i.e. to abort, and that the fetus is nothing else than a "bunch of cells".

The expected suffering and early death of the newborn baby is Mrs B's reason to abort. However, Mrs B also reacts against the doctor's expression that the fetus is just a "bunch of cells". She thinks it is a child; "this is my child", she says and she wants to bury the aborted fetus. Hence, from the interview it is obvious that Mrs B has established a relationship with the fetus.

Mrs B's moral view can be interpreted in the following way: The fetus is a child with human dignity. Nevertheless, due to the expected suffering and death of the newborn, it is better to abort the fetus than to let it live a short and painful life. Hence, the application of the principle of human dignity – to let the fetus live – is restricted by the fetus' expected suffering and early death. Based on this way of reasoning one can *specify* the principle of human dignity in the following way: according to the principle of human dignity the fetus has a right to live, *unless* it will suffer and die soon after birth (for specification, see Richardson 2000).

Is then Mrs B facing a moral dilemma? From one point of view, Mrs B seems to face a moral dilemma having to choose between the life and death of the fetus. On the other hand, she avoids the choice. Perhaps one can describe the situation in the following way: Mrs B can avoid making a choice by taking for granted her doctor's recommendation, that a fetus with Edwards' syndrome should be aborted. Hence, she does not need to take a decision.

13.3.3 Interview C

Mrs C is 36 years old and 8 months pregnant. She sees her pregnancy as a "great present". Due to her age she is offered amniocenteses but she refuses. The reason is that she is worried that the test will threaten the life of the fetus. Her partner agreed and said he was able to cope with a disabled child and her parents and parents-in-law also supported her.

The interview with Mrs C is an interesting case of someone who intuitively decides not to have PND alongside the psychological and ethical problems connected to this decision. It raises many important ethical questions.

What are the implications of a decision to decline PND? Mrs C decides – apparently without doubt and on a well-informed basis – not to have PND, although she is of an age when the risk to the fetus is increased and to have the test seems to be what society expects of her. This implies – as is mirrored in the interview – that she is responsible for the possible outcome that a disabled baby could be born. What does this imply in terms of her relation to the baby and to society? And, on the other hand, what would it imply if she had PND against her own moral conviction?

Thus, the interview also highlights the ethical implications connected to societies' "offer" of PND. Although PND is voluntary, it might be experienced as almost compulsory by pregnant women over a certain age. Is such an offer consistent with a principle of respect for autonomy, emphasising that each person has the right to decide him/herself about matters of his or her own concern? Is the sense of compulsion perhaps in conflict with this important principle? On the other hand, what is the ethical implication of the fact that another individual is involved, namely the fetus? Who decides on behalf of the fetus?

13.3.4 Interview D

This interview focuses, in the main, on Mrs D's reasoning regarding PND and whether to terminate the pregnancy or not. It says something about her present social situation but very little about her life story.

Mrs D has two daughters. She underwent PND because of her age (over 30). One of her daughters has a genetic disorder. From the interview, it is not clear how prenatal testing affects her decision concerning her pregnancy. She has had a previous abortion and a miscarriage just before the present pregnancy. Mrs D is a medical doctor and she knows a lot about the different methods of prenatal diagnosis. Her husband is in the army and plays a minor role in the decision making.

The interview does not reveal what decision Mrs D makes on the basis of genetic information. However, one can notice that she stresses that it is *her* decision and that neither family – including her husband – nor friends have any influence over it. However, one factor that influences Mrs D's decision is that she has a daughter with a minor genetic disorder (see below).

Two factors seem to be important in order to understand Mrs D's reasoning. One is her striving for security. She wants to know everything about her pregnancy and the status of the fetus. She has ultrasound tests every week and she is almost obsessed by facts and figures about the probabilities of genetic disorders etc.

The second factor is the fact that she has a daughter with a genetic disorder (*Congenital adrenal hyperplasia*). The disorder causes early sexual development and androgen tendencies like hairiness and ovarian cysts. The daughter has two mutations, one from Mrs D and one from her husband. As both parents are carriers this disorder might appear in future children. How does this fact influence her decision? On the one hand it seems that she is worried by the thought of her next child being a carrier. On the other hand, in conversation with her daughter, she maintains that this disorder would not be a reason to abort. For her, terminating a pregnancy, based on information about the fetus carrying the daughter's genetic disorder, would be similar to terminating her daughter.

The interview does not reveal what decision Mrs D took on the basis of PND, therefore we do not know her reasoning concerning the question of whether to abort or not. However, we do have some information from the interview. On the one hand she states that genetic defects are "for life" and that having such a child "can destroy a whole family". On the other hand, although her preoccupation with testing and security might be interpreted in another way, she denies that any disorder within the fetus would be reason to terminate a pregnancy. She considers the fetus to be a potential "amazing child".

For Mrs D, the fact that she has a daughter with a genetic disorder has important moral implications for the decision whether to abort the fetus or not. She explains that to abort the fetus because it carries the same genetic disorder that her daughter has, would be equivalent to "abort", i.e. to kill, her daughter.

Mrs D's argument, based on a connection between having a daughter that she loves and the fetus, can be understood in at least two ways. First, she would not accept that any disorder within the fetus is a reason for abortion. She seems to think that her daughter's life, although affected by the genetic disorder, is a valuable life. Secondly, she establishes a relationship between her daughter's life and the question of whether to terminate the fetus or not. How can this relationship be established and what is its moral meaning? One possible line of thought is the following: The daughter lives with the genetic disorder. It affects her personality making her more masculine. The androgen-linked effects are balanced by steroids. However, even if this genetic disorder causes difficulties, it is part of her personality. Her identity is at least partly shaped by her genetic make-up. Without it she would have been another person. But it is this person that is the daughter that the mother loves. Hence, if the genetic disorder is a reason to abort the fetus, it implies that a life with this genetic disorder is not worth living. While the genetic disorder is an integral part of the daughter's identity, the decision to abort would imply that the daughter's life is not valuable.

The availability of different kinds of genetic testing invites an attitude of full certainty that is mirrored in the interview with Mrs D. However, one may ask, is this attitude morally sound? First, it does not seem to correspond to a realistic worldview. Despite prenatal testing, there is always a degree of uncertainty and a lack of knowledge. Secondly, is the need for absolute certainty a character trait that is desirable? How much control is ethically justified?

Mrs D's attitude towards PND appears to be contradictory. On the one hand she is almost obsessed by a need to control every moment of the pregnancy. She also acknowledges abortion based on PND when she states that a child that is born with a genetic disorder can destroy a whole family. On the other hand, she would not accept any reason for terminating the pregnancy. Hence, she explicitly argues against abortion in the specific case of a genetic disorder. In this way her intimate relation with her daughter has implications for her decision whether to terminate pregnancy or not.

13.4 Interpretations of Interviews

13.4.1 *Explanation*

We have so far given account of four interviews with mothers facing the decision to abort or not based on PND. How can a mother's reasoning and decision-making be explained? To ask for explanation means to ask for causal relations. What caused the reasoning and decision? What factors were important as background factors?

The interviews show that two factors are of particular importance. One is the social relations of the prospective parents. There are people close to the mother, like a partner, parents or brothers and sisters, who give advice or in other ways influence decision-making.

Another important factor behind the decision is the mother's earlier experiences. These experiences are of different kinds: of a death in the family, of having a disabled brother or sister, of having encountered people with illnesses or abnormalities, earlier abortions etc.

When finding explanations for a decision to keep or to abort the fetus, it is important to distinguish between subjective and objective explanations. Subjective explanations refer to the mother's personal explanations of her decision. She may express views about which people are important to her decision and about what earlier experiences have been influential. However, the mother's story may or may not give a true or full explanation of whom or what in reality influenced the decision. Of course, while it is the mothers' stories that are the empirical basis for this analysis, the interviews can only provide information for subjective explanations.

13.4.2 Understanding

To understand a decision is to see it as meaningful. A decision is in this sense meaningful as an integrated element of a life story or personal narrative. In an analysis that focuses on the mothers' decision-making, it is the mother's own story that gives us information about the meaning of a decision to terminate a pregnancy or not. A psychoanalysis may reveal how a particular decision is meaningful as part of a life story.

In order to understand a mother's story we would focus on what personal experiences she emphasises in her story? Was she free to take any decision or was she bound or restricted in any way? Does she mention any religious denomination or ideological group that might have influenced her decision? The answer to this last question gives us (partial) information about her sense of identity. As Charles Taylor explains, a person's identity is not only revealed through his or her stand on moral and spiritual matters but also with reference to a "defining community" (Taylor 1989, p. 36).

The interviews referred to above give us some indications for understanding the mothers' decisions. For example, we can better understand a mother's decision when we know that she already has a daughter with a genetic disorder which has shaped her life or that a mother has had a brother with a disorder which stamped the life of her family when she grew up.

13.4.3 Justification

What is the prudential or moral underpinning of a decision to keep or abort the fetus? An example of a prudential justification of a decision to abort is that the mother is not able to take care of a child with a disability. It is assumed that an abortion will imply a less burdened life. An example of a moral justification is when a mother refers to an ethical principle, for example the principle of human dignity, as reason not to abort the fetus.

Moral considerations are of course factors that are important for justification. The mother may for example refer to a norm or an ethical principle to justify a decision. Other examples of moral considerations are references to intrinsic values and a view of a good life. For example, the mother may argue that life with a disability is not worth living, thus, abortion of a fetus with a genetic disorder is morally justified. In the interviews we came across some examples of moral reasoning.

One mother referred to the possible suffering of a disabled child as a reason for abortion and another mother wanted to bury the aborted fetus, presumably as an act of reverence for the dignity of the unborn child.

Another kind of justification refers to commitment and identification. For example, the mother is committed to an important job and hence she can not take care of a disabled child. Another example: the mother identifies with her own mother who took care of a disabled child. Thus, she wants herself to keep the fetus even if it is disabled. In one of the interviews a mother revealed a commitment to a daughter who was carrier of a genetic disease.

Also relevant for the question of justification is whether the woman regrets her decision or not. If a decision is regretted, the justification is shaken.

13.5 Conclusion

The four selected interviews inform us about the mothers' experiences, reflections and personal views. What, then, are the implications of this kind of information for ethics? Ethics is usually understood as the theory of morality, i.e. the rational deliberation of moral thinking and decision-making. However, narratives told by individuals faced with difficult decisions, for example after a positive test result after prenatal diagnosis, complement theoretical insights. As Eve-Marie Engels writes: "…individual narratives can bridge the gap between real life and theoretical ethics" (Engels 2008, p. 270). Then, this kind of information is highly relevant for an applied ethics that not only strives to be theoretically well-informed but also wants to make a contribution to moral practice.

References

Engels EM (2008) Experience and ethics: ethical and methodological reflections on the integration of the EDIG study in the ethical landscape. In: Leuzinger-Bohleber M, Engels EM, Tsiantis J (eds.) The janus face of prenatal diagnostics: a European study bridging ethics, psychoanalysis, and medicine. Karnac, London, pp 251–272

Hildt E (2008) Moral dilemmas and decision-making in prenatal genetic testing. In: Leuzinger-Bohleber M, Engels EM, Tsiantis J (eds.) The janus face of prenatal diagnostics: a European study bridging ethics, psychoanalysis, and medicine. Karnac, London, pp 273–287

Leuzinger-Bohleber M, Engels E-M, Tsiantis J (eds.) (2008) The janus face of prenatal diagnostics: a European study bridging ethics, psychoanalysis, and medicine. Karnac, London

MacIntyre A (1981) After virtue: a study in moral theory, 2nd edn. Duckworth, London

Medicine, Health Care and Philosophy (2004) Empirical ethics (special issue). Med Health Care Philos 7(1):1–134

Richardson H (2000) Specifying, balancing, and interpreting bioethical principles. J Med Philos 25(3):285–307

Taylor Ch (1989) Sources of the self: the making of the modern identity. Harvard University Press, Cambridge

Chapter 14
Prenatal Diagnostics and Ethical Dilemmas in a Mother Having a Child with Down Syndrome

Maria Samakouri, Evgenia Tsatalmpasidou, Konstantia Ladopoulou, Magdalini Katsikidou, Miltos Livaditis, and Nicolas Tzavaras

Abstract Prenatal diagnosis is supposed to allow autonomous decision making on the basis of personal goals, plans and values. However, often this type of decision-making may not be achieved, as various factors can be implicated in choosing an alternative between the available options concerning a pregnancy. The following case illustration presents a woman who decides against an amniocentesis, although there is an increased risk for a positive finding in prenatal diagnostics (PND), due to her age and a genetic problem in her family history. The aim of presenting this case is to show how such a decision may be associated with individual factors in one's life. The clinical observations from the European, interdisciplinary EDIG study, which investigated how individuals experience ethical dilemmas due to prenatal diagnostics, suggest that certain behavioural characteristic (i.e. expressing ambivalent feelings and thoughts) and/or important past life experiences (i.e. affirming serious losses in one's history) may comprise indicators for risk and protective factors for the psychological processes of women/couples facing a problematic situation when undergoing PND. The following case, Mrs K., participated in the EDIG study via responding to a series of questionnaires and two interviews. Both interviews were discussed within research group meetings and were qualitatively analysed. Analysis further supported the EDIG study's hypotheses regarding the ways in which the manifestation of certain personality traits and past life experiences, may influence responses to decisions after PND. Important implications for medical and counselling practice were also provided.

Keywords Case study • Down syndrome • Prenatal diagnostics • Psychoanalysis

---

M. Samakouri (✉), E. Tsatalmpasidou, M. Katsikidou,
M. Livaditis, and N. Tzavaras
Department of Psychiatry, Democritus University of Thrace,
68100 Alexandroupolis, Greece
e-mail: msamakou@med.duth.gr

K. Ladopoulou
Children's Mental Health Centre of Athens, Child Psychiatry Hospital of Attika,
4 Garefi, 11525 Athens, Greece

14.1 Introduction

While analysing qualitative data obtained during the "Ethical Dilemmas due to Prenatal and Genetic Diagnostics" (EDIG) study and investigating how individuals experience ethical dilemmas due to prenatal diagnostics (PND), it has been observed that individual and biographical characteristics of one's own inner psychic world may be involved in coping with the possibility of giving birth to a disabled child and the necessity for life or death decisions (Leuzinger-Bohleber et al. 2008).

The case we present here is about a woman who was willing to be interviewed for the purposes of the EDIG project and who, despite the increased risk of a positive PND finding due to her age and a genetic problem in her family history, decides against an amniocentesis and chooses to continue her pregnancy, no matter what "fate will bring".

The aim of presenting this case illustration is to demonstrate that, even if one decides not to undergo PND, a choice which may not appear to present obvious ethical dilemmas, such a decision may be associated with idiosyncratic factors. According to one of the main hypotheses formulated for the purposes of the EDIG project, the way in which women/couples undergoing PND present their life experiences and their personality characteristics may be indicators for risk or protective factors, as far as psychic functioning associated with decision-making, is concerned (Leuzinger-Bohleber et al. 2008).

Mrs K. participated in the EDIG study, completing a series of questionnaires and two interviews conducted at two different time points during the project. The two semi-structured interviews were carried out by a psychiatrist, a member of the Thrace research group in Greece, and according to the EDIG study research protocol instructions for interviews. Both interviews were discussed within the group meetings for the purposes of qualitative analyses.

14.2 Case Illustration

Mrs K. was approached when she was reaching the 23rd week of her sixth pregnancy by a member of our research team, in the waiting room of an obstetric private practice and whilst she was waiting for her appointment with her doctor to undergo a second trimester ultrasound scan (anomaly scan). She had already been informed by the obstetrician of our study and she seemed enthusiastic to participate. That was the first strong emotional reaction, which characterized her behaviour and was registered as a special transference condition.

At the time, Mrs K. was a 42-year old mother of five children (two girls – 19 and 14 years old – and three boys – 15, 10 and 8 years old). Eight years ago Mrs K. had to face a difficult situation, as her last child was born with Down syndrome, a condition which was not diagnosed during her pregnancy and which she found out about only after birth.

14.2.1 The First Interview

At the first interview, which takes place after Mrs K. has already undergone an anomaly scan, she is three quarters of an hour late. She is a very talkative and pleasant woman. From the beginning of the interview she talks about her present unplanned pregnancy and her doubts about it, although she clearly states her negative attitude towards an abortion. Yet, she does not explain the reasons for her delay, giving the impression that her expressiveness may cover up possible inner conflicts.

> For me a pregnancy was always a joy… But now I am more anxious, as I am very occupied with raising my last child. I was not very pleased with this pregnancy. Yet, I have never thought of having an abortion. I have never had one in my life and I would not consider having one now.

Soon after, Mrs K. chooses to focus on all the possible advantages associated with the arrival of her new baby. One can see that all the pros of having a baby are mainly related to the fact that her 8 year old son has Down syndrome.

> … I thought of all the possible positive things I could get out of having this baby. First of all, I will take maternity leave from my job. I won't be working for a while and I need this time for my son … I also want to have some rest, since I was very tired over the last period working with him when he started going to school… Then I thought that this would be very good for my son, as my other children are older and the one we are waiting for, will help him… he will have some company in the house.

At this point in the interview, Mrs K. unfolds, along with her maternal love, the rigidity of her Superego, which is projected on the child she will give birth to. This child should also – like herself – 'meet' the needs of his/her older disabled brother. Mrs K. gives the sense that she unconsciously aims to become ethically acceptable to the interviewer, since she refers to the efforts she makes in bringing up her disabled child.

Mrs K. is an accountant and she has been married for about 20 years to a 52 year-old civil engineer. She reports that she and her husband are still very much in love with each other after so many years and that, despite some everyday arguments and complaints, they have a very close relationship. Yet, for the first time her husband is quite uncertain about her pregnancy, being afraid that the baby may be born with an abnormality. For the first time, he suggests to his wife that she has an abortion – a suggestion that he then regrets *"not for the sake of her reaction but rather because he never liked the idea of an abortion anyway"*. Mrs K. perceives his initial reaction as a moment of weakness and fear.

When asked about the support she receives from her husband, she mentions that he had been very supportive in practical matters (i.e. helping the children to study, going for shopping, cooking at weekends), although in the last years he has held back and does not take over many things in the house; something which Mrs K. blames herself for.

> When I first gave birth to my child with Down syndrome, I felt that there may be something wrong with me. That is, that I am not able to have a healthy child. I felt that I was in a vulnerable position and guilty, in the sense that I was the one insisting on having a fifth

child. My husband wanted it but he was suggesting we should wait for two more years. Anyway... he finally agreed. But the truth is that I forced him. And once our child was born with the syndrome, I felt that I had given him a child that he may not have wanted or that he would have not had with another woman.

When considering the birth of their son with Down syndrome, Mrs K. reveals her contradictory feelings. Following the unexpected birth of a disabled child, her optimism that she could overcome all difficulties (as determined by her Superego) is replaced by a sense of guilt towards her husband. At the same time, her low self-confidence and particularly the challenge to her physical, maternal adequacy, emerge. It is also obvious that her almost hypo-manically coloured hopefulness is superseded by an at least transient, depressive reaction (Freud 1916–1917).

Respectively, the counter-transference emotions initially obtain a negative quality, as the interviewer identifies herself with the members of Mrs K.'s family, whose needs and potentials – mostly those of her husband – are tested from her wish to continuously procreate. Mrs K. becomes even more likeable, once she transfers her despair to the interviewer who facilitates the discussion. As far as the counter-transference reactions are concerned, it seems as though the interviewer's identifications falter between Mrs K. and the members of the family (Sandler et al. 1970).

She agrees with the interviewer's impression that after giving birth to her last child it was as though she was reacting as if she did not have any healthy children and that this was a burden to her.

> It affected me and influenced the relationship with my husband. I was indifferent and inaccessible and, due to my guilt, I took over too many things, so I would not tire him and from that time my husband stopped helping me. He started being more relaxed and relying on me handling many things at the same time.

However, things have now changed in the way she feels. She does not feel guilty any more. In contrast, she thinks she has helped her child, who is doing very well now, a lot and that makes her feel proud of herself. This also enables her to give advice to other women who are in the same position, in terms of what they should or should not do.

In her previous pregnancy, Mrs K. did not consider the possibility that something may go wrong. She only had an anomaly scan and her doctor reassured her that everything was fine. It was after birth, when she called the doctor that did the ultrasound, who told her that only 30% of Down syndrome cases are diagnosed in the anomaly scan and those are usually serious cases.

In an attempt to work alone on her guilt feelings and her fantasies of her ethical and physical incompetence, Mrs K. is led towards a burden including new obligations that may ease the qualms of her conscience. It is about a strategy of an ethical self-reassurance, which in parallel obscures her own reality. The interviewer's fantasies – as well as those of the group members that jointly discussed the case, are directed towards the rest of Mrs K.'s children that appear 'abandoned' (Green 1965).

> "And as it was proven" she says "our child is not a serious case; he is totally healthy. He has not got heart problems that could have been observed by the ultrasound."

14 Prenatal Diagnostics and Ethical Dilemmas

That is the reason, she thinks, she can talk easily about her experience with her child, as if she had a more serious problem she might have seen things differently.

> I am in a better position than mothers who do not understand the condition and who may think it is tragic. It is not tragic. That is why my telephone number is available from my doctor to all mothers who may have a positive PND finding. If a mother wants to see me, to talk with me, if she wants to decide she can come and see my child... Because that is the way it is really. You can do miracles with this child and you get much more than you will get from other children. He makes us proud when we are watching him achieving things that are difficult for him to achieve. You can see that when trying this little child can do many things and he is making us very proud. I cannot say I am embarrassed about my kid; I am very proud of him.

Mrs K. idealizes her young boy. This obviously supports her efforts to convince herself of her ethical supremacy, as well as to induce the admiration of her narcissistic wholeness, used to dispel all possible within-family difficulties.

Mrs. K. also thinks that at the time the amniocentesis takes place, it is very late and *"You have already felt the baby inside you. It is a child in shape. You have seen it on the ultrasound."*

That is not easy for her, since if this baby has an abnormality, she cannot tell the effect that this may have on her family. Yet, she considers abortion as murder; *"I think you are killing it. It is a murder!"* she states. Nevertheless, she would not reject the possibility of a pregnancy termination in the very first weeks following conception, if she could know that there was a very serious condition. But it would have to be a more serious condition than Down syndrome.

> Living with this child and making the most out of it, I do not consider it as a problem any more.

Interestingly though, she chooses to have a second trimester ultrasound (anomaly scan) in her current pregnancy, mainly because having the reassurance of a negative test result with a 95–98% certainty, giving her peace of mind and making her happier and more optimistic.

Some of the most characteristic information Mrs K. gives, when talking about having the anomaly scan, is that her children – apart from the youngest one – were all present at the examination. She vividly describes that scene, emphasizing their great concern and enthusiasm – particularly that of her oldest 15-year old son – to see the baby.

Mrs K.'s oldest son is presented as the most anxious and responsible of all of her children. He was often asking her during the pregnancy what is the possibility of this baby being born with a genetic disorder. When she refers to her oldest daughter she chooses to talk about her attitude towards a pregnancy termination, which is similar to hers, both for fetal abnormality and an unplanned pregnancy.

Mrs K. directly embroils her children in the problems she is preoccupied with. Her avoidance of asking herself whether the children do themselves wish to be presented with the inside of the maternal body and to encounter the problems she is confronted with, is very characteristic. The atmosphere established, since the father is more or less absent, has an incestuous dimension, according to which, once again, the mother demands that her children accept her decisions, her ethical attitude and her physical destiny (von Klitzing and Burgin 2005). In the counter-transference, the pictures obviously carried over and the interview seems to bring about queries and a sense of disaffection. Once again, the interviewer's associations comprise the question of what the older children would have wished for. The interviewer's associational thinking revolves around the objects-children that the mother makes, throughout the conversation, her self-objects.

Over the last 14 years, following the birth of their third child, Mrs and Mr K. became very close to the Greek Orthodox religion. For her, she explains that feeling so happy inside her family she thought she had to thank God. Despite her strong religious affiliations, she insists that her attitude towards PND and abortion does not stem solely from her religious beliefs, but rises also from an ethical point of view.

Although feeling obliged by her husband, doctors, and most friends to do all the tests required, one person that supports her decision not to undergo amniocentesis, is her mother, who is described as a strong person who seems to interfere a lot in Mrs K.'s family's life, usually in a negative way. Her mother lost her husband, who suffered from a chronic disease, 5 years ago and her youngest son in a car accident when Mrs K. was 24. The death of Mrs K.'s brother is underplayed in this discussion. Once asked about whether this loss has affected her views about life, she shifts her attention onto other events, such as a recent diagnosis of a breast cancer in one of her friends, had affected her views about life. Nevertheless, as to her brother's death she maintains it is something she has forgotten.

When talking in more detail about her decision not to undergo amniocentesis, Mrs K. remembers her doctor talking about the possibility of a positive finding and about her responsibility to make a decision. It is that remark which makes her decide against an amniocentesis. On that day, she recalls being so scared and experiencing physical symptoms (her heart rate increases and she is about to faint). She also starts to remember her last labour and she re-experiences the birth of her disabled child. It is at that point in the interview that she reports a very horrible fantasy, which occurred when her doctor explained to her that for an abortion to take place after the 20th week, obstetricians administer medication to the woman to kill the embryo and afterwards they cause physical labour.

> The fantasy I had was bad. To give birth to a dead child. That was the fantasy I had. To give birth to a baby knowing that I wanted to kill it. That is, to think about myself pushing with hate to get this dead baby out of me. To force myself, with my own will, to get it out of my womb, whilst it was too early for the baby to be born. That was my fantasy, which shocked me.

The interview ends with her accepting only helpful aspects of PND. According to Mrs K., if a positive finding occurs this is not negative, despite the dilemma it may bring. At least you are prepared and know what to do about it. And if a woman decides to have an abortion, she would not blame her.

> Everyone has the right to decide on her own for her life and her body.

During the second part of the interview, the discussion with Mrs K. is characterized by a dramatic tension, introduced through the image she creates once her children attend her ultrasound examination. Immediately after that, her Superego demands emerge, throughout her mentioning of the role the Orthodox Church plays in her life. Her faith is expressed through her gratefulness to God. Indeed, Mrs K. denies that her views about pregnancy termination are dictated by religion, but rather they owe their existence to her personal, ethical appraisal, implying that religiousness does not shape her ethical rigidity. Mrs K.'s personal ethical claim has actually been determined by her mother's sadistic behaviour (Green 1965).

Stories of deaths and illnesses follow: the death of the father 5 years ago, the death of the brother, the cancer of a friend. Thus, Mrs K. reveals her inner psychic reality, which consists of a series of internalized fatal objects. She talks about her horrific fantasies concerning the pregnancy with a dead child. In this way, Mrs K. expresses her ambivalence stemming from a conflict of an essential significance between her aggressiveness towards pregnancy and her submission to the demands of a sadistic introjected mother. It seems that all the relative ambivalent feelings she suppresses, may also concern her brother's death, which by that time has remained without any emotive after-effects (Green 1965).

14.2.2 *Further Points Discussed Within the Research Group*

During the first interview with Mrs K., a sense of admiration is created within the group. Everyone is quite impressed by her highly expressive way of talking and the smooth language she uses. Nevertheless, this impression is soon replaced by feelings of anger and dysphoria, once the group distinguishes the presence of some omnipotent feelings and ideas of self-grandiosity. She gives the impression of a woman who tries to admonish others. She represents the mother that can help the child with the Down syndrome make very good progress. This is also evident from the way she overestimates the capabilities of her child with Down syndrome. Such thoughts are also indicated in the way she rejects scientific progress and describes the help she offers to parents who are facing a dilemma. Omnipotent feelings, as hypothesised in the group, may be related to the guilt feelings she may have experienced for the loss of her brother.

On the other hand, it is very characteristic that she expresses guilt feelings, for giving birth to a child with an abnormality, especially towards her husband. That makes the group think of the possibility that she may want to overcome feelings of inadequacy and inferiority by turning into the ideas of self grandiosity. Although once she felt she was not capable of giving birth to a healthy child, she is nevertheless, the mother who can take over many responsibilities, be very optimistic, help her child, feel proud about herself and become able to advise others. It has also been hypothesized within the group's discussion that bringing her children into her ultrasound examination might have been an attempt to reassure her family. It might have been a process of undoing and becoming the mother who can have a healthy child inside her womb.

Regarding the death of her brother, the initial suspicion of the group is that there may be some unconscious guilt feelings. Such feelings, as we have already emphasized, could stem from former aggressive experiences focusing mainly on her brother.

The reference to that relationship, near the end of the interview – though its content is initially obscured – leads Mrs K. to convey, through associations, a series of facts concerning death. Mrs K. avoids talking about this loss in more detail and chooses to refer to other experiences, which might have affected her life. The group begins to wonder whether her negative attitude towards pregnancy termination is unconsciously connected to the condition of her son and the loss of her brother. The feelings of extreme happiness expressed by Mrs K. may also serve as defensive purposes against that painful experience which in this interview is underplayed (Freud 1957).

Although Mrs K. clearly states her negative attitude towards abortion, it is still quite shocking to her, once asked by her doctor what she would do in case of a pathological finding. The physical symptoms and the stress she experiences, as well as her fantasy of giving birth to a dead child, may be considered as indicators of her not being capable of facing a dilemma and experiencing guilt feelings. It seems possible that she experiences an inner strain: she cannot bear a positive diagnosis. That is why, by deciding not to have an amniocentesis, she pretends it does not exist. She'd rather be in a state of a relative certainty rather than have a dilemma – a denial mechanism.

Somehow unexpectedly, Mrs K. makes the interviewer become the communicant of her terrifying fantasies related to her own fatal wishes. At this point, her hypo-manic and 'optimistic' defence is replaced by the revelation of her aggressive ambivalence. It is the interviewer's presence that facilitates this really impressive reversal perceived as a request for a further psychotherapeutic session.

A transgenerational phenomenon, in terms of the attitudes shared between mother and daughter and which passed on to the granddaughter, is also observed. Mrs K.'s daughter becomes the agent of her ethics, the same way she became the agent of her mother's ideas (Lebovici 1988).

The ethical dilemma is presented in relation to the personality characteristics of Mrs K. During the validation, the counter-transference reactions of the group members are mainly related to her Superego rigidity. Mrs K.'s self-idealization serving mainly as defence purposes of the unconscious sides of her Superego, appears through her missionary tendencies and the dispute of scientific procedures in the evaluation of a pregnancy. Mrs K. does not accept any challenge in the course of a pregnancy and stresses the happiness originating from extended motherhood. Nevertheless, at around the end of the interview, this convention starts acquiring a different character, particularly once the issue of Mrs K.'s repressed aggression making use of more straight-forward language (thanks to the protective presence of the interviewer), appears.

14.2.3 The Second Interview

The interviewer meets Mrs. K. about 4 months after the birth of her sixth child, a healthy boy. She immediately starts to explain how the long labour was encountered as a nice experience, emphasizing her great relief when she saw that the baby was healthy. The thought of further childbirth is a stressful one. Yet, she does not reject this possibility, repeating once again her negative attitude towards abortions.

Mrs. K. always liked being a mother of many children.

> The idea of not having another child was always disturbing to me. I was feeling sorry for not being able to re-experience having a child, but I knew I could not afford it and on the other hand, I was thinking I could not be able to love that child, since I became barren by giving myself to my son.

But this time Mrs. K. was having a difficult time. About a month following labour, she reports going through an overwhelming psychological situation, including symptoms of "*melancholia, depression and lots of obsessional ideas*".

The interviewee expresses herself at a greater ease, within the frame of this second interview, and over the description of her psychological states. One gets the impression that she is expecting to benefit from the interview – to talk about the things that have happened between the first and this second meeting. In contrast to the previous interview whereby Mrs K. was presented as elated and grandiose, probably due to manic defences through which she used to dispel her daily existential difficulties, she is now led to a depressive state which she decides to talk about.

> I had a very hard time. Watching the news on TV I was wondering 'How will my child be in 50 years time?' 'Is there going to be any water on the earth?' 'Will there be a disaster?'. I used to listen to a heretic speaker on the radio talking about the end of the world. For a whole month I was sitting up all night flooded with anxiety and fear. Once it was getting dark outside, I was feeling scared.

Her fantasies about the end of the world released her from her multiple obligations, bringing about not only her destruction, but also the catastrophe of all the incriminatory objects that she loves and hates at the same time. Her wish to withdraw from the world is projected to the external reality, as a destruction of great magnitude. Her unconscious disastrous tendencies are disguised in a delirious way to an inevitable physical disaster, which she is not responsible for.

She admits that during that month she did not want to take care of herself, her husband and her children. She also tries to give an explanation by recalling that she was not resting enough over that period, since she was breastfeeding and was having home visitors. She had always had a constant fear of the end of the world and the antichrist, but she had never been scared for so long. She asked for the help of her confessor. Following confession, those thoughts stopped occurring. Yet, they were replaced by a new "obsession", as she says.

> I was convinced that my husband's mind was with another woman… I thought there was something going on with our next-door neighbour who broke up with her boyfriend and used to discuss this with my husband, when he used to smoke on our balcony… I thought he used to live more on the balcony than inside our house.

When the fantasies of a whole world disaster receded, another paranoid fantasy of a milder form appeared – Mrs K. attributes to her husband a love interest towards another woman. The pictures of his discussions with their female neighbour ensure the excitation of the primal scene. This keeps her away from the horror of the disaster

(Pedrina 2006). It nevertheless enables her to distance herself from the annoying objects and to take the central role in a love 'drama'. The group that elaborates and comments on the interview interprets the search for the primal scene, as an unconscious, particular strategy of the interviewee to regress and to avoid her resumption of her daily routine, with a new child (Freud 1916–1917).

In terms of her son with Down syndrome, Mrs. K. is still taking care of him, although not so much as she used to.

> I don't think I have neglected him very much. Well, maybe I did a bit. I feel that my caring is not the same and that this is better as before I used to force him with his studies and his behaviour in general. Now, I am more relaxed and I can handle things in a better way. I even started realizing that some things that used to annoy me are due to his condition... I think he performs better at school now.

Although she notices that her little son is quite jealous of the baby, the two will soon share a bedroom. In this way, she believes the child with Down syndrome will be happier and less jealous.

At the end of this discussion, the topic of social support arises. Mrs. K. takes the opportunity to talk about the practical help she receives from her mother (associated with child rearing), but she also states she cannot count on her emotional understanding and support.

Mrs K.'s husband on the other hand, was very helpful and supportive after the last child was born. However, she thinks that this was not the case during the period she was not feeling well. Their relationship has been very good for many years. But over the last years they argue a lot mainly in terms of his behaviour towards their oldest daughter, who has recently entered University. Mrs. K. thinks of her husband as a very strict and irritable person, yet a very good husband and father. She is worried that being restrictive to their daughter may damage her trust towards him. She recalls memories from the past when she used to feel very disappointed by her parents being quite restrictive towards her.

It is at this point of the interview where more signs of emotions are evident with respect to Mrs K.'s past. The sudden death of Mrs. K's brother seems to have affected her in many ways. One important aspect may be associated with her decision to have many children.

> The truth is that I feel very lonely without my brother. I feel that I cannot share things of my childhood. That is why I decided to have many children. So as they will have one another. Although I always wanted it, once he died I was surer than ever... My mother also used to have a brother who was 4 years younger than her, as my brother was to me. She lost her brother in an accident, too. And I was wondering whether the history is repeated. Thus, I said I will have many children so I can change the course of the history.

Mrs K. expresses very differently the emotions of herself and those of the members of her family. The interviewer is very impressed by the links she can recall between her early experiences and her present life; particularly, how the traumas from the past determine her maternal efforts. Through the latter, she wants to expel the traces of her trauma. Her brother is reborn through the birth of every new child, whilst at the same time, the repression of all negative feelings associated with him, is perpetuated. Fantasies of hatred described during the first interview, implied the intensity of all those negative feelings. A pregnancy termination would mean for Mrs K. the recurrence of her favourite brother's death, for which she considers herself responsible. She believes that his death is nevertheless not associated with her attitude towards abortions and to her fears for the end of the world, as she had those much earlier. After that loss she becomes more familiar with death and she is not afraid of it anymore. Yet, she reports experiencing phobias related to the possible loss of her family members at different times of her life.

> In the past I was thinking that someone in my family will die; one of my children or my husband. I experienced that very intensely… I was anxious about car accidents. That was after my brother's death. I used to say 'I d rather die first as I will not come to terms with something like that'.

The loss of Mrs K.'s brother may also seem to have influenced her attitudes towards PND.

> I do not absolutely believe in amniocentesis, since you may receive a negative finding, and your baby is born with autism or something goes wrong during labour and it gets an encephalic paralysis. I believe that everything is a matter of luck…

The interviewer's question as to whether she may need professional support and help in the future is denied by Mrs. K. She does not accept that psychiatrists could help her. She always used to take the advice of her confessor and he has been the one who helped her over the last months, too.

14.2.4 Further Points Discussed Within the Research Group

From the beginning, there is an impression that Mrs K. is more open to disclose more of her emotions, in this second interview.

The research group estimates that Mrs K. feels unable to deal with all the responsibilities emerging from the extended motherhood. The development of her symptoms may signify a realization on her side of her inadequacy to meet huge obligations. In this sense her depression, as well as her feelings of jealousy towards her husband, which appear afterwards, may be related to her need to pull back from the work she took over as a mother of many children. On the other hand, it seems that her newborn baby allows her to cope well with the condition of her son and may serve as a reparative psychological function.

During the second interview Mrs. K. decides to talk about her feelings towards her brother's death. The initial connections made by the research group between self idealization, death fantasies, guilt feelings, and the initial resistance to talk about this loss during the first interview prove to be correct in terms of her not working through this traumatic event. This event becomes even more unbearable in considering its tragic transgenerational repetition. When coming to terms with a shared traumatic experience with her mother, the history becomes destiny (Lebovici 1988).

It seems that this loss has affected Mrs. K. in many ways: Firstly, she states clearly that her brother's loss has affected her in her decision to have many children in order for them to have one another. It seems as if giving birth to many children may help her work out the negative and guilt feelings she has towards her brother. It is also a way to fight death and disrupt the tragic destiny within her family. Moreover, Mrs K. occasionally admits experiencing phobias associated with the loss of her family members. She is particularly afraid of car accidents. Such phobias may also be considered as indicating the effects of her brother's loss on her life. One may also assume there is a connection with her turning to religion. Mrs K. wants to thank God, as she states in the first interview, for her happiness. Or is it because her children are still alive? Further, it is discussed within the group, that in an indirect way this loss influences her attitude to PND. She believes everything is a matter of luck in terms of developing a health problem, in an attempt to justify her decision against an amniocentesis. The interviewer thinks that having experienced the misfortune of losses of young persons in the family, this view is to be expected.

Based on their emotional reactions, the members of the group see again the great transitions made during the two interviews. Initially, counter-transference emotions of admiration and rejection (due to her hypomanic exaggeration) gave the impression that Mrs K. avoids the emergence of her inner conflicts. Yet, the course of the discussions reveal differentiated aspects of her personality, potentialities of verbalization and the presence of an inner dialogue. An atmosphere of disappointment is caused amongst the members of the team, once the interviewee states she was not willing to be further supported psychotherapeutically. For a short while, feelings of rivalry towards her 'confessor' – who seems to play a reassuring role with respect to her constant feelings of owing and guilt – , predominate.

14.3 Conclusion

In conclusion, it seems that the death of Mrs K.'s brother being experienced as a traumatic event, along with the transgenerational family experience of early losses, may have strongly influenced her attitude towards prenatal diagnostics and her conscious decision not to undergo an amniocentesis, but rather to continue a high-risk pregnancy. Through the first interview, Mrs K. is characterized by a mild grandiosity. She praises herself and she presents as being ethical. This certainty is, nevertheless, disrupted by the possibility of a positive diagnosis and the thought of death. During the second interview Mrs K. is more able to express her ambivalence in herself and to better understand some of her inner conflicts. In line with the clinical observations, based on the interview analyses within the EDIG study, the present case also illustrates that defining risk factors associated with the psychic functioning of people coming to terms with decisions evoked by PND is of great importance. Mrs K.'s early losses and guilt feelings that had to be repressed seemed to have influenced her attitude towards life and death decisions. In addition, providing an empathetic, professional dialogue – as that which seemed to unfold through the interviews with Mrs K. who finally dealt with her traumatic experiences in a more mature way, may be helpful for individuals experiencing analogous conditions

(Leuzinger-Bohleber et al. 2008). On the basis of these assumptions, training medical and paramedical staff a) to identify the individuals who may need more specialised professional help and b) to provide appropriate information in a supportive way, as well as improving counselling practice in the field of PND, can be considered important aims for the future.

References

Freud S (1916–1917) Vorlesungen zur Einführung in die Psychoanalyse. Band XI GW
Freud S (1957) Mourning and melancholia: the standard edition of the complete psychological works of Sigmund Freud, vol 14. Hogarth Press and the Institute of Psychoanalysis, London
Green A (1965) Private madness. Hogarth, London
Lebovici S (1988) Interaction fantasmatique et transmission intergénérationelle. In: Cramer B (ed.) Psychiatrie du bébé. Eshel, Paris, p 321335
Leuzinger-Bohleber M, Belz A, Caverzasi E, Fischmann T, Hau S, Tsiantis J, Tzavaras N (2008) Interviewing women and couples after prenatal and genetic diagnostics. In: Leuzinger-Bohleber M, Engels E-M, Tsiantis J (eds.) The Janus face of prenatal diagnostics: a European study bridging ethics, psychoanalysis, and medicine. Karnac Books Ltd., London, pp 151–218
Pedrina E (2006) Mütter und Babys in psychischen Krisen. Forschungsstudie in einer therapeutisch geleiteten Mutter-Säugling-Gruppe am Beispiel postpartaler Depression. Brandes und Apsel, Frankfurt
Sandler J, Holder A, Dare C (1970) Basic psychoanalytic concepts: IV Counter-transference. B J Psych 117:83–88
von Klitzing K, Burgin D (2005) Parental capacities for triadic relationships during pregnancy: early predictors of children's behavioral and representational functioning at preschool age. Infant Mental Health J 26(1):19–39

Chapter 15
Is There One Way of Looking at Ethical Dilemmas in Different Cultures?

Stephan Hau

Abstract A continuous public debate on ethical questions in the field of prenatal diagnostics is seen as necessary. However, it is not self-evident that such a discussion is taking place. One reason for this complication can be seen in the different cultures of discourse within a society, representing different interests. This article examines two different "cultures" of discourse: a scientific discourse of experts and a public discourse of lay people. Different aspects and problems were addressed. However, the more complete picture comprising all the various perspectives is only achieved when multiple discourses are taken into consideration. Finding platforms for exchange seems to be a continuous problem even though new media technologies offer better opportunities for communication.

Keywords Prenatal diagnostics • Public debate • Scientific debate

In times of globalisation one can often read about the impact of different cultures within a society. In this context one might usually think about individuals from different countries, having different cultural and migrational backgrounds or showing different religious beliefs. These and other differences can be understood as important components underpinning the morals and social structures of social groups as well as an individual's personality and belief system. It is not selfgiven that one automatically and easily understands these cultural backgrounds. However, knowledge about these different cultural contexts is essential. They become relevant in connection to conflicts, disputes, controversial discussions, and ethical dilemmas.

Taking this into consideration it seems quite easy to answer the initial question of this article: There cannot be a single way to look at and explore ethical dilemmas within a society consisting of different cultures, even when confronted with the

---

S. Hau (✉)
Psykologiska institutionen, Stockholm University, 106 91, Stockholm, Sweden
e-mail: stephan.hau@psychology.su.se

same dilemmas. It is of great importance to focus on cultural differences within a country like Sweden that has accepted and has been in the process of integrating thousands of migrants in recent years (e.g. since the war in Iraq began, more refugees and immigrants from Iraq live in the small Swedish city of Södertälje than have been admitted to the United States and Canada).

When different cultures are involved, a typical example of the difficulties of these discussions can be observed when considering the process of introducing a new test method in the field of prenatal diagnostics. However, there is another issue that will be highlighted and investigated in this article: that of representing the discourses of different cultures *within* a society. That is investigating the discrepancies and differences between a scientific discourse about a specific topic and the public debate about the very same issue: the former tries to compile and assess the state of the art of expert knowledge, the latter takes place in the media (e.g. in newspaper comments, internet blogs) and contains many fears and irrational fantasies. This discourse contains important objections that are often excluded in the scientific discourse. One can assume that different levels of involvement are represented in these debates reflecting different "cultures" within a society; in this case the cultures of scientific research and evidence on the one hand and the culture of personal, and often subjectively biased, involvement on the other hand. It is quite clear these different contingents of expert knowledge, fantasies, beliefs, and interpretations shape the various comments. To get a better picture on how ethical dilemmas are handled differently in different cultures within a society the relevant discourses have to be explored more thoroughly.

A general problem in Sweden had been that no public debate took place when decisions were made to introduce new methods of genetic prenatal diagnoses. It was the practitioners who made the decision, believing that women would want better tests. Any ethical considerations were of minor relevance. As a consequence, the rapid decision to introduce a new method made it almost impossible to halt it, e.g. for the reason that time might be needed for further ethical discussion. This situation forced the Swedish Council on Medical Ethics (SMER) to investigate the problem in order to publish an ethical opinion. This investigation was done independently of the other project described below performed by the Swedish Council on Technology Assessment in Health Care (SUB). As a result SMER stated the need to monitor ongoing ethical conflicts which should also include the opinions of lay people (e.g. patients, patient organizations, politicians). It became clear that a permanent platform would be needed in order to guarantee continuous discussions.

In a recent study (Dekker 2009) the public debate which took place about prenatal diagnoses between 1989 and 2006 was summarised. As found by the SMER review, one of the results was that public debate in Sweden had been dominated by experts. Interestingly, Dekker noticed that traditional media (e.g. radio, newspapers) are not suitable for this public debate. The internet is seen as a good resource to initiate increased participation among lay people.

In what follows two examples of debates/dialogues are described representing different "cultures" of debate within society.

One example of the scientific debate is a report by the "Swedish Council on Technology Assessment in Health Care (SBU)"[1] (Nilsson et al. 2006) which evaluated thoroughly the current methods of early prenatal diagnostics. The second example is based on a systematic media analysis of major articles in the two leading Swedish newspapers "Dagens Nyheter" and "Svenska Dagbladet" from November 2007 until January 2008 including readers' reactions and comments on the same topic.[2] In accordance with Dekker's findings (2009) the introduction of a combined ultrasound and biochemical test led to an intense debate among lay people on the internet.[3] Neither a comprehensive overview was intended nor to comment in a judgemental way.

In contrasting these discourses, it becomes apparent that not all relevant arguments are represented in each discourse. Thus, when exploring ethical dilemmas across different cultures it seems necessary to channel a more systematic exchange of different debates.

15.1 The Culture of Scientific Discourse

15.1.1 Background

When in 2007 the "KUB-Test" (Swedish abbreviation for "*k*ombinerat *u*ltraljud och *b*iokemi test – combined ultrasound and biochemical test") was introduced as a non-invasive screening method offered to all women in the first trimester of pregnancy (in the County of Stockholm) the number of amniocenteses decreased by 30%, the time between test and test results shrank to a couple of days, and about 30% of all pregnant women chose the KUB-test (Dagens Nyheter [DN], 25.11.2007). At the same time, an intense public debate started about the advantages and disadvantages of the test. In particular, the debate in the media raised ethical questions concerning the attitude towards chromosomal abnormalities, e.g. Down syndrome, about the right of women to abort, and about the "hunt for abnormalities" in prenatal diagnostics.

It is remarkable that this debate takes place in a country with a liberal tradition, giving pregnant women the right to chose freely whether they want to have an

---

[1]SBU (Statens beredning för medicinsk utvärdering) is a governmental authority (www.sbu.se) carrying out the commission to evaluate methods applied in the medical services on a scientific basis for all relevant decision makers in the field.

[2]This is of course not a statistical analysis or a representative study. The comments were only categorised (e.g. positive or negative) and then summarised according to different contents addressed.

[3]Interestingly, the debate has stopped again. No comments or statements can be found on the homepages of Dagens Nyheter or Svenska Dagbladet in 2009.

abortion or not.[4] The debate cannot only be understood as an example of how deeply prenatal diagnostics affect society and the public debate on health care but illustrates the complexity of the dilemma and that no "perfect" solutions can be achieved.

15.1.2 The Scientific Approach

In 2006 a detailed report by the "Swedish Council on Technology Assessment in Health Care" was published (Nilsson et al. 2006) in which currently available methods of early prenatal diagnostics were evaluated thoroughly. The conclusions formulated in the report were based on research findings and had a strong impact on how to organise and finance the prenatal diagnostic tests which should be available for all pregnant women in the country.

A combined test (ultrasound and biochemical serum test), carried out during weeks 10–14 of the pregnancy, seemed to show the best results when investigating the risk for Down syndrome (i.e. the relation between correctly identified cases and wrongly positive results) (Nilsson et al. 2006, p. 17). An important conclusion was that, above all, ethical considerations have to be taken into account when introducing the KUB-test as a routine test for all pregnant women as well as the other important economical, social, and psychological implications. With respect to prenatal diagnostics, shortcomings were described such as different motives for testing, adequacy of information, lack of knowledge, economical consequences, and quality control.

15.1.3 Motives for Testing

Pregnant women and doctors want to achieve the same: early and safe testing. However, there is a significant difference in their perspectives of the aims of genetic prenatal diagnostics. For most of the women, it is of priority to know that everything is normal with the fetus while professionals, in the medical sector, aim to find and diagnose abnormalities. (Nilsson et al. 2006, p. 512; Farsides et al. 2004; Williams et al. 2002a, b). This creates an area of tension that makes adequate information processes for women essential for their decisions, especially to ensure that there should be no causal connection between offering prenatal diagnostics and abortion (Nilsson et al. 2006, p. 514; Vetenskapsradet 2001).

---

[4]In 1973 a liberal law legalising abortions was implemented in Sweden. Basically women got the right to decide for abortion without giving any special reasons. In the preceding years an intense discussion about gender roles was going on in the country.

15.1.4 Information

The process of information giving as well as the judgements of risks is seen as inadequate. Ideally, information should be given orally and in writing by a specialised medical doctor or by a geneticist. In practice, midwives mainly deliver the information verbally in about 5–10 min (Nilsson et al. 2006, p. 517, 522; Cederholm et al. 1999). This leads to a limited basis of knowledge for the women to make their decisions. Furthermore, it cannot be excluded that there are differences in how the information is given and genetic diagnostics are offered, depending on the socio-economical status of the patients or their ethnic background. Even though ethnocultural sensitivity and empathy seems to be particularly relevant in this context (Rasoal 2009) no indepth research exists in Sweden which clearly describes the current situation.

About one third of women seek additional sources of information (e.g. from friends, books, or journals). Diagnosis of abnormities is seen as the most important overall reason for PND, but there are also others e.g. to determine the sex of the fetus. About one third of the women could not remember whether they were informed whether abnormalities were found (Nilsson et al. 2006, p. 517; Bricker et al. 2000; Eurenius et al. 1997; Crang-Svalenius et al. 1996).

The last point suggests the importance of psychological aspects that should not be underestimated: in a stressful situation with a lot of anxiety, even the clearest information may not be fully understood but unconsciously altered or misinterpreted or neglected. Special knowledge is necessary about the psychological mechanisms for coping with stressors. Information models have to take these into consideration in order not to widen the gap between the different "cultures".

Furthermore, there seem to be significant differences between the clinics performing prenatal diagnostics. Better education of professionals (staff) is needed and more money must be invested not only in better equipment but also in having more time available to inform the women and their partners (Nilsson et al. 2006, p. 587). This is also important because there is an increased level of worries and anxieties among women who do not feel well informed. Thus more knowledge may be associated with decreased anxiety level. Women's concerns after amniocentesis is performed relate more to the possibility of miscarriage than to giving birth to a disabled child (Nilsson et al. 2006, p. 512). Comprehensive information seems to be a possible way to minimize stress and anxiety.

There are some basic aspects that should be addressed when information is given. The relevance of either a positive or negative result should be explained as well what possible choices for further actions exist. In Sweden, these demands cannot be seen as fulfilled (Nilsson et al. 2006, p. 587; Green et al. 2004). In addition, information is delivered differently in various parts of the country, and in particular clinics, because the professionals' attitudes influence the way they talk to the women, even though the law requires the same general information for all women. Therefore, midwives and other medical staff need more and better education in ways of presenting objective information and about the ethical questions that are implied. In addition, we learn that differences exist not only due to the different cultures from which the women are coming but also due to the ethos in the different clinics and the settings, i.e. when information is given and by whom.

15.1.5 Lack of Knowledge

There is no Swedish consensus about the best way this information should be delivered (Nilsson et al. 2006, p. 521). In general a lack of knowledge about PND can be observed, especially when it comes to the large number of women with a non-Swedish cultural and ethnic background. Moreover, there are no studies about the knowledge of midwives and medical doctors, their attitudes and how their education might be improved, nor studies focussing on women's partners.

What is completely missing is research on long-term effects of genetic diagnostics on the psychological health of women, on their attitudes towards the child, on the results of different methods in PND as well as on short and long-term effects and on quality control (Nilsson et al. 2006, p. 527).

15.1.6 Control of Quality

Quality control is seen as essential for a well functioning system of genetic diagnostics. In general, there is a lack of quality control for all parts of the process of genetic diagnostics as well as for different clinics. There seem to be no adequate models for quality control on how prenatal diagnostics is performed. This seems to be crucial when considering that there might be different information procedures necessary depending on the knowledge of the women, their cultural backgrounds and their ethical beliefs.[5]

Even though many different aspects are addressed in this report, there is a clear limit to "objective" data and to technical and management approaches. An individualised approach, taking the individuals, their psychologies and cultural differences into consideration is not described. Of course, one has to be aware of the purpose of this report, but it is still remarkable that the individuals' perspectives do not appear.

---

[5]In the UK quality controls/standards for its PND services exist – www.fetalanomaly.screening.nhs.uk

15.2 The Public Debate – Another Culture of Discourse in a Modern Society

The process of introducing the KUB-Test in Stockholm county as an additional diagnostic/screening possibility for all women led to an intense public debate in the Swedish media. It was the first time that an intensive debate among lay people had taken place. Some of the main arguments are summarised here.

In the data set are major articles and readers' comments that were published within a period of 3 months (November 2007 until January 2008) in the printed newspapers and on the homepages of the newspapers "Dagens Nyheter" and "Svenska Dagbladet". This is, by no means, a representative compilation or a statistical analysis of data. However, all comments were categorised (positive or negative comments, different contents addressed). Far more negative than positive reactions were published and the statements included a higher level of emotional involvement than the publications of the scientific debate. Finally, arguments concerning abortion or PND were also contained in the readers' comments. It became obvious that many fears, fantasies, and emotionally laden thoughts are a major part of this discourse.

15.2.1 Discussions About Abortion

The increase in the numbers of abortions of fetuses with abnormalities as a consequence of better genetic diagnostics is seen as a danger that would undermine the basic humanistic idea of equality of all human beings (Sandberg 2007). It is assumed that the scientists do not see the ethical problems they create when focusing eagerly on better diagnostic tools. They could be either blind to the consequences of what they are doing or there might be an assumption of a hidden agenda which is not overtly expressed. In this case, this would mean implicitly favouring abortion as the most likely choice after positive PND-findings (Sandberg 2007).

Cultural as well as private beliefs and attitudes – without being officially acknowledged – are seen as strong motives that influence societal developments. Depending on beliefs and values in a society, the fear is that eugenics can secretly or unintentionally develop, as in China, which is mentioned as an example, where males are more valued than females (Linusson 2007). Most recently, the debate about the Swedish regulation of abortion has started again when a woman decided to have an abortion after receiving the information about the sex of the fetus. She had already had previous abortions because the child was the "wrong" sex. As a consequence, the social authorities have now been asked to develop recommendations for new regulations.

As if this subject was not difficult enough, another question complicates the debate. From which point in pregnancy can one assume the fetus to be a real "human being"? Lagercrantz (2008) lists a line of arguments that tries to describe an essential difference between a fetus on the one hand, clearly seen not as a human being, and a baby after birth on the other hand. The crucial criterion described by him is the moment of onset of conscious mental activity, which is seen as given as soon as the baby is able to breathe (independently or in the incubator). From this moment, parents would be able to develop and experience a "me-you" relation to the baby that from that point has started to develop a soul (Lagercrantz 2008).

As a consequence of this logic it would be possible to abort fetuses with chromosomal abnormalities as well as healthy fetuses before week 23 of pregnancy. Lagercrantz (2008) thinks the conclusion is skewed: one does not violate human dignity by aborting fetuses with severe abnormalities before week 23, but at the same time, one can also abort healthy fetuses because of their lack of human dignity before the onset of having a soul.

One point that catches attention is the simple idea of mental capacity and of the circumstances when conscious experiences are given. Due to this approach, unconscious mental activity is excluded as well as precursors of conscious experiences. Furthermore the assumption regarding fantasies that mothers and fathers have about their unborn children are untenable. Parents will not make their fantasies and expectations about the fetus/child dependent on any breathing activity. Unconscious as well as conscious fantasies already exist long before birth, sometimes even before conception. These fantasies generate a specific relational situation, e.g. depending to a great extent on the relational and attachment experiences of the parents but also on cultural factors.

15.2.2 Unwanted Abnormalities

Another sensitive subject is the demand to discuss and determine which abnormalities are seen as unwanted and where it is permissible to interrupt a pregnancy. Linusson (2007) believes that the technical development of better testing may give rise to a "new fascism", e.g. against human beings with Down syndrome. They would be "cleaned away" eugenically by being declared as unwanted humans. This "cleansing" would be attempted as effectively and as early as possible by the new developments in prenatal diagnosis, like offering the KUB-Test to every pregnant woman. If this tendency continues to develop there is speculation about what might be the next step of defining unwanted diagnoses or criteria: "left-handedness, homosexuality, corpulence, nearsightedness, intelligence, or small stature?" (Linusson 2007, transl.: S. Hau).

These arguments seem to be irrational and completely exaggerated; however, they mirror anxieties and fears that are within the individuals.

Another controversial aspect of the debate is to assume adequate and sufficient information for all pregnant women. If relatively simple information like vaccination against cervical cancer does not reach the people who need to know, how much more problematic must it be to build up adequate knowledge about the KUB-test, prenatal testing and its consequences (comment, SvD)?

Besides these critical voices, positive statements underline that the KUB test is less dangerous compared to amniocentesis (with 1% risk of causing miscarriages) and can be performed earlier in pregnancy. The method could cut down the numbers of amniocenteses by 50% and at the same time detect three times as many chromosomal abnormalities than before (Rydberg 2007). The mean age of women becoming pregnant for the first time is 30 years (Stockholm), and in other regions of Sweden even higher, and rising. Many women want the KUB-test even today. The official claim is that there is no automatic abortion after a positive finding. The institution would have to be neutral and the decision would be made by the individual woman. However, there are differences within the country, e.g. in the southern parts fewer tests are performed (Rydberg 2007).

In addition, moderate contributions to the discussion evaluate the current situation positively. Equal access for everyone, independent of income, cultural background or knowledge would exist. Generally, it is underlined that the KUB-test creates an opportunity to discover chromosomal abnormalities in a better and more precise way. Thus, the risk of false positive results can be reduced and neutral and competent information would be proposed, including details about the implications of having a child with Down syndrome, about possible help and about how people with Down syndrome can live a good life (Rydberg 2007).

While parents have the right to dream about their children, society cannot define which children are wonderful and which ones are not (Linusson 2008). However, there would be a bias in public health focussing on research for tests about diagnosing Down syndrome and not putting effort into studies which would broaden our knowledge about the best ways to prepare pregnant women for their difficult decision. This situation would portray an implicit message: society is expecting a decision for abortion if the fetus has Down syndrome.

This view elicits critical comments. Even if information were provided for women it could not necessarily be assumed that the medical staff can present the correct information in a competent way. Linusson (2007) states that some of the leading medical doctors wrongly believe that the KUB test can replace all amniocenteses. He ends in a rather shrill conclusion: If decision makers like politicians or the leading medical personal are that uninformed with respect to crucial important ethical question "the warning bells should ring from Europe's darkest fascist history" (Linusson 2007). The fear is that a society will develop a very problematic view on differences, even though the development towards a human society would be in need of difference, a crucial cornerstone for an open, human, and democratic society (comment, SvD).

Individual parents cannot be blamed if they don't want to have a child with Down syndrome. However, the grade and quality of information upon which their decision is based would often be tremendously low. Irrational fantasies exist about how horrible it would be to have a child with Down syndrome. As authorities pay for a test that can detect these deviations, parents would arrive at the conclusion that is must be horrible to give birth to and raise a child with Down syndrome (commenter SvD, Linusson 2007).

Parents with children with Down syndrome took part in the debate. Their comments describe the shock of the diagnosis on the one hand and on the other hand the gladness and joy after birth when relating to and experiencing the child. The following case might illustrate this as an example:

When Mrs. X found that her newborn daughter might have Down syndrome she was in emotional chaos. Today the daughter is 2.5 years old and Mrs. X and her husband see her as "a wonderful gift of God". Before the birth of her daughter Mrs. X, who has a good education, a satisfying job, and lives in a self-renovated house hoped to have a "perfect" family: herself, her husband and two daughters. The child came 4 weeks early and the only thing Mrs. X reacted to was her short fingers. The doctors decided to hospitalise the child in the incubator assuming the diagnosis of Down syndrome. "I felt that I was in a deep, dark hell. It was especially bad because I had prenatal diagnosis. The test somehow promoted fears and anxiety about Down syndrome. A Down-child would have been the worst that might hit our family."

The parents are convinced that their daughter has an unbelievably strong will to live a happy life because she had managed to trick herself through the system. "In the beginning we just saw a heavy burden of care to carry. Now she is a part of the family. And for us a new world has started, a new dimension has opened. We are thinking new thoughts. We meet new people. And our situation creates new thoughts and experiences."

Now, 2 ½ years later the mother feels glad and happy and the daughter is a bright, clever, sweet little child. Today the parents have completely new insights. Their shock and anxiety has changed to happiness about their youngest daughter. Of course, it took longer for the daughter to learn to speak and to learn to walk. However, with extra effort from the parents she learned the same things as other children. The parents are most critical of the fact that they were not well informed about Down syndrome, how to deal with these children and how they grow. In addition, they are critical of the tendency to detect children with Down syndrome prenatally, in order to abort them (commenter SvD).

Some commentators suspect that the public enthusiasm for prenatal diagnostics is only helpful for the researchers themselves. However, it would be more important to talk with those who live with a disabled child, sibling, grandchild, cousin or friend in order to receive more realistic information.

It seems difficult for expectant parents to develop a positive perspective about having such a child when research tries to prevent as many births as possible of these "unwanted children". Between the lines of all information and reports there is the suggestion that abnormalities have to be avoided whenever it is possible.

However, the picture has many facets. The following example shows how early ultrasound tests can be a good alternative and have a reassuring effect for the parents. In this case the couple belonged to a part of the population, which found it difficult to become parents due to age and miscarriage. They belong to a group with high risk for a disabled child (over 1% risk for Down syndrome). On the other hand invasive prenatal diagnostic has a risk of miscarriage (about 1% as well). That was a "horrible situation" for them, especially when they learnt of others who had found it difficult to have children and then suffered miscarriage because of invasive diagnostic testing.

15 Is There One Way of Looking at Ethical Dilemmas in Different Cultures? 201

It was a positive experience that a very experienced person performed an ultrasound test. When they learned that the risk of having a child with Down syndrome was less than one in 500, this seemed to be a very low risk compared to having a miscarriage because of invasive testing. They felt relaxed during the rest of the pregnancy. Today they have a healthy child (commenter SvD).

15.3 Some Conclusions

The examples given should illustrate the complexity of the different dimensions that are affected and play a role in the context of ethical dilemmas. The case here was the introduction of a new method in prenatal diagnostics. Different cultures do not necessarily mean different ethnic groups within a society. A scientific and a public discourse can be understood as different clusters of debates representing different aspects of coping with a dilemma. It seems to be necessary to create exchanges in the debates in order to achieve a broader picture and to capture the complexities of such a process.

In the end a dilemma cannot be solved like a crisis or a conflict. That is why there is a tendency to revive older dilemmas in the context of new discussions. In the example given there seems to be a tendency in the public debate to confound two different fields of controversies: discussion about prenatal diagnostics and about abortion.

Some shortcomings in the debates within the different cultures can be identified. In the scientific report psychological dimensions are severely neglected and the importance of the necessity of working through such a stressful situation is underestimated. Especially after a "shocking" experience, more sensitivity to the need for psychological help and support of the women and their partners could be developed.

What is also missing are studies about long-term effects of PND. The results of the EDIG project are, in that respect, of special importance. The focus so far has been on the short time period before and after the diagnostic procedure. That there is a need to work through these experiences can be seen, to some extent, by the popular blogs which exist on the Internet, in which parents vividly exchange their experiences and "talk" about them in detail.

Another striking aspect of the debate is that specific contents, effects or implicit tendencies involved in the field of prenatal diagnoses are not addressed. This aspect is brought up not by the scientists involved but by the participants of the debate in the newspapers, another "culture" so to speak. As far as the public debate is concerned, several types of fears and anxieties turn out to be active in individuals which are not seen and captured by the scientific summaries. This demonstrates the importance of transparency and of a culture of discourse in a democratic society in order to balance the different shortcomings.

From a psychoanalytic point of view what is missing, or not seriously taken into consideration, is the dimension of unconscious processes. It is well known that on an individual level unconscious content permanently influence the consciously experienced psychological state. Examples can be found in several chapters of this book. Furthermore, psychoanalytic researchers have demonstrated that even on group levels, institutional levels and on a society level trends that lay out of conscious range play an important influential role (cf. Mentzos 1988). Here, the participants of the debate in Sweden gave several hints that give rise to further investigations.

To conclude, even in a highly developed medical care system, considerable shortcomings can be found, that become apparent in different cultures of discourse and that ask for careful and responsible discussions as well as actions.

References

Bricker L, Garcia J, Henderson J, Mugford M, Neilson J, Roberts T et al (2000) Ultrasound screening in pregnancy: a systematic review of the clinical effectiveness, cost-effectiveness and women's views. Health Technol Assess 4(16):1–193

Cederholm M, Axelsson O, Sjoden P (1999) Women's knowledge, concerns and psychological reactions before undergoing an invasive procedure for prenatal karyotyping. Ultrasound Obstet Gynecol 14:267–272

Crang-Svalenius E, Dykes A, Jorgensen C (1996) Organized routine ultrasound in the second trimester – one hundred women's experiences. J Matern Fetal Invest 6:219–222

Dagens Nyheter 25.11.2007: Gravida ska erbjudas metod mäta kromosomfel

Dekker C (2009) Judging in the public realm: a Kantian approach to the deliberative concept of ethico-political judgment and an inquiry into public discourse on prenatal diagnosis. Linköping University Electronic Press, Linköping. http://urn.kb.se/resolve?urn=urn:nbn:se:liu:diva-16942

Eurenius K, Axelsson O, Gallstedt-Fransson I, Sjoden P (1997) Perception of information, expectations and experiences among women and their partners attending a second-trimester routine ultrasound scan. Ultrasound Obstet Gynecol 9:86–90

Farsides B, Williams C, Alderson P (2004) Aiming towards "moral equilibrium": health care professionals' views on working within the morally contested field of antenatal screening. J Med Ethics 30:505–509

Green J, Hewison J, Bekker H, Bryant L, Cuckle H (2004) Psychosocial aspects of genetic screening of pregnant women and their newborns: a systematic review. Health Techn Assess 8(iii): ix–x, 1–109

Lagercrantz H (2008) Människovärde först vid födseln. Svenska Dagbladet, 3 Jan 2008

Leuzinger-Bohleber M, Engels E-M, Tsiantis J (eds.) (2008) The Janus face of prenatal diagnostics: a European study bridging ethics, psychoanalysis, and medicine. Karnac, London

Linusson S (2007) Nyfascism kring Downs syndrom. Svenska Dagbladet, 28 Nov 2007

Linusson S (2008) Vi vet vad som är ett underbart barn. Svenska Dagbladet, 3 Jan 2008

Mentzos S (1988) Interpersonale und institutionalisierte Abwehr. Suhrkamp, Frankfurt

Nilsson K, Alton V, Axelsson O, Bokström H, Bui T-H, Crang-Svalenius E, Eckerlund I, Eksell S, Lindgren P, Löfmark R, Maršál K, Saltvedt S, Tunón K, Valentin L (2006) Metoder för tidig fosterdiagnostik (Report No. 182). The Swedish Council on Technology Assessment in Health Care, Stockholm

Rasoal C (2009) Ethnocultural empathy. Linköping studies in arts and science No. 505, Linköping University, Department of Behavioural Sciences and Learning

Rydberg B (2007) Kromosomtest leder inte automatiskt till abort. Svenska Dagbladet, 29 Nov 2007

Sandberg N-E (2007) Är livet alltid okränkbart? Svenska Dagbladet, 28 Dec 2007

Svenska Dagbladets hompage (comments). http://www.svd.se/opinion/brannpunkt/. Accessed 20 Feb 2008

Vetenskapsradet (2001) Tidig fosterdiagnostik – Konsensusuttalande; Konsensuskonferens 10–12 okt 2001 i samverkan mellan Landstingsförbundet, Socialstyrelsen och Vetenskapsrådet; ISBN 91-7307-004-1. Vetenskapsrådet, Stockholm

Wahlström J (2009) Prenatal diagnoses. Paper presented at the European Group on Ethics in Science and New Technologies meeting, Stockholm, 15–18 September 2009. http://ec.europa.eu/european_group_ethics/activities/. Accessed 03 Feb 2010

Williams C, Alderson P, Farsides B (2002a) Dilemmas encountered by health practitioners offering nuchal translucency screening: a qualitative case study. Prenat Diagn 22:216–220

Williams C, Alderson P, Farsides B (2002b) What constitutes "balanced" information in the practitioners' portrayals of Down's syndrome? Midwifery 18:230–237

Index

A
Abortion, 2, 4, 5, 9, 14, 18–21, 57, 60–63, 78, 81, 84–86, 105, 110–113, 122, 125, 132, 136, 140, 142, 144, 145, 149, 156–159, 161, 169, 170, 172–175, 179, 181, 182, 184, 185, 187, 194, 197–199, 201
Alpha fetoprotein (AFP), 54, 160
Amniocentesis, 3, 12, 17, 20, 52, 55, 67, 68, 73, 76, 78, 79, 81, 84, 85, 90, 122, 124, 126, 132, 133, 135, 169, 170, 178, 181, 182, 184, 187, 188, 195, 199
Antenatal care, 8, 85, 88, 155
Anxiety, 24, 53–55, 59, 61, 62, 66, 72, 73, 114, 126, 132, 135, 185, 195, 200
Archaic, 7–9, 23–27, 29–31
Aristotle, 155–160, 162–164
Autonomy, 21, 30, 61, 62, 112, 114, 116, 123, 131, 148, 171

B
Bereavement, 6, 13, 86, 92
Bundesärztekammer, 101, 104

C
Catholic, 4, 111
Child, 3, 7, 9, 12–26, 28, 31, 32, 54, 55, 57, 59–62, 68–73, 78–82, 86, 87, 91, 92, 100–101, 105, 111, 113, 117, 121, 125, 127–136, 140, 141, 144, 149, 150, 157–163, 167, 169–175, 177–189, 195–201
Childhood, 7, 16, 18, 20, 21, 26, 28, 69, 134, 169, 186
Chorionic villus sampling (CVS), 19, 52, 90, 100
Christian, 4, 57, 60
Coping, 2, 3, 6, 11, 15, 17, 21–24, 26, 27, 29, 30, 54, 55, 59, 61–63, 84, 132, 178, 195, 201
Counselling, 9, 27, 44, 52, 61–63, 66, 85, 88, 96, 97, 100–107, 122–124, 127, 130, 132, 133, 136, 139–153, 159, 189
Crisis intervention, 44, 52
Culture, 2, 4, 32, 84, 85, 149, 152, 191–202

D
Database, 43
Death, 2, 3, 7–9, 14, 15, 18, 20, 22–26, 32, 55, 60, 61, 78, 79, 86, 125, 129, 141, 147, 157, 170–171, 173, 178, 182–184, 186–188
Decision, 2–4, 12, 14–20, 22, 23, 25, 29, 32, 46, 53–56, 60–63, 70, 71, 76, 78, 79, 84, 85, 87, 89–92, 94, 100, 110–112, 114, 115, 117, 118, 122–125, 127, 130, 132, 134, 136, 140, 141, 144, 147–150, 153, 155–164, 168–175, 178, 182, 186–188, 192, 199
Decision-making, 5–6, 15, 37, 56, 61–63, 75–79, 81, 82, 84, 88, 90–91, 93, 96, 100, 106, 110, 111, 114–116, 123–128, 132, 133, 136, 147–149, 155–160, 162–164, 167–175, 178
Depression, 3, 7, 9, 25, 51, 53–55, 62, 114, 185, 187
Dilemma, 80–82, 122, 123, 148, 150, 153, 157
Disability, 13, 24, 57, 62, 70, 71, 125, 130–132, 134–136, 139, 141, 143, 160, 162, 174
Dissemination, 4, 38, 41–42, 48

Index

Distress, 2, 3, 5, 13, 14, 51–64, 66, 73, 86, 110, 114–117, 128, 130, 146, 167, 169
Down syndrome, 5, 54, 77, 84, 87, 117, 124, 125, 127, 131, 132, 141, 149, 159, 160, 177–189, 193, 194, 198–200

E

Embryo, 21, 81, 82, 158, 161, 182
Emotions, 6, 7, 9, 12, 13, 20, 23, 27–30, 64, 130, 132–133, 135, 148, 180, 186–188
Ethical, 4, 9, 31, 37, 53, 56–60, 62, 76, 82, 88, 92, 93, 96–97, 100, 107, 109, 110, 112, 113, 115–119, 122, 123, 129, 130, 132–134, 136, 139–146, 149, 150, 152, 153, 156, 168, 169, 171, 173, 174, 179–183, 188, 192–194, 196, 197, 199
Ethical dilemma, 2–4, 9, 12, 20, 22, 39, 43, 44, 46, 52, 54, 56–58, 100, 135, 150, 177–189, 191–201
Ethical dilemmas due to prenatal and genetic diagnostics (EDIG), 1–32, 35–48, 51–64, 66, 76–77, 80–81, 83, 84, 88–92, 109, 110, 112–114, 118, 126, 128, 130, 132, 134, 143, 147, 148, 150, 151, 156, 157, 160, 168, 169, 178, 188, 201
Expert validation, 11, 17, 22, 54

F

Fantasies, 6–9, 14, 23–25, 27, 28, 31, 37, 60, 66–69, 106, 128, 180, 183–187, 192, 197–199
Fertility, 7
Fetal abnormality, 5, 56, 75, 85, 90, 160, 181
Fetus, 2, 3, 8, 9, 12, 14, 15, 24, 58–62, 75–79, 81, 82, 84, 111, 113, 122, 124, 125, 127–131, 134–136, 140–144, 147–149, 158, 161, 167–175, 194, 195, 197–199

G

Genetic counselling, 63, 85, 100, 109–119, 122, 159
Genetic diagnosis, 170
Genetic diagnostics, 167–175, 195–197
Genetic research, 2, 96
Grief, 5, 6, 9, 64, 92
Guilt, 5, 7–9, 13, 14, 20, 21, 24–27, 29, 55, 61, 62, 78, 114, 145, 159, 179, 180, 183, 184, 187, 188

H

Haemophilia, 18, 20, 21, 84, 134, 135, 169, 170
Human dignity, 58, 59, 81, 105, 113, 170, 171, 174, 198

I

Interruption of pregnancy, 17, 22–26
Interviewees, 22, 169, 185, 186, 188
Interviews, 2, 4, 9, 11–22, 24, 27, 31, 39, 40, 46, 63, 76, 94, 103, 105, 107, 134, 149, 150, 168–175, 178, 188
Intrinsic values, 174
Islamic, 60

J

Jewish, 60, 62

M

Malformed, 9
Mapping technique, 35–48
Medea fantasy, 7, 8, 24
Mind map, 42–48
Moral, 2, 56, 60, 78, 79, 81, 82, 113, 121–136, 144, 147–150, 152, 166–175, 191
Moral dilemma, 2, 150, 151, 168, 170, 171
Motherhood, 6, 8, 79, 184, 187
Mourning, 8, 13, 17, 26, 29, 55, 62
Mr. E., 16–17, 32
Mrs. C., 12–14
Mrs. E., 15–17
Mrs. F., 18–21, 23, 31, 134

N

Narcissism, 9
Narcissistic, 8, 24, 26, 30, 31
Narrative, 132, 168–170, 174
Nationaler Ethikrat, 4

P

Paranoid-schizoid position, 23, 27
Perinatal loss, 6
Pregnancy, 2, 5–9, 12–17, 19, 20, 22–26, 53–58, 61–62, 66, 69, 71–73, 75–81, 84–94, 100, 102, 105–107, 110, 111, 113–116, 118, 121, 122, 124–129, 131, 132, 134–136, 140–142, 145, 149, 152, 155–161, 163, 167, 170–174, 178–181, 183, 184, 187, 188, 193, 194, 197–199, 201

Pregnant, 3, 6, 8, 13–15, 18–21, 25, 39, 51, 52, 54, 55, 59, 61, 65–73, 75, 76, 79–81, 83, 84, 86–88, 93, 97, 101, 104–107, 121–136, 140–144, 146, 156–164, 169, 171, 192, 193, 198, 199
Prenatal and genetic diagnostics, 1–32, 51–73, 85, 99–107, 122, 126, 129, 130, 135, 139, 140, 156, 158, 162, 167–175, 177–189, 192–197, 200, 201
Prenatal diagnosis (PND), 2–9, 12, 14, 15, 18, 21–32, 37, 39, 40, 44, 46, 51–54, 56–63, 65–73, 76, 79, 83–97, 99–107, 121–126, 128–130, 132–136, 139–146, 156, 158, 162, 167, 168, 171–173, 175, 177–189, 192–198, 200, 201
Prenatal screening, 5, 76, 84, 88, 100, 109, 117, 160
Professionals, 2–4, 9, 11, 13–15, 18, 21, 23, 26–28, 30, 32, 61, 63, 73, 76, 84–90, 92–97, 100–105, 107, 109–111, 114–116, 118, 122–124, 126, 128, 130, 133–136, 141, 144, 148, 158, 187–189, 194, 195
Protective factor, 2, 3, 9, 22, 26–30, 64, 178
Psychoanalysis, 7, 18–25, 37, 40, 52, 56, 149–152, 174
Psychoanalyst, 2, 4, 6, 7, 9–11, 23, 24, 26, 37, 39, 40, 42, 48, 63, 64, 117, 128, 146, 150, 151
Psychological distress, 5, 73, 116, 128, 130
Psychopathology, 2

Q

Qualitative content analysis, 67
Questionnaire, 2, 4, 45, 51, 53, 55, 56, 63, 66, 68, 70, 76–80, 82, 126, 128–132, 156–158, 161, 163, 178

R

Regression, 9, 24–26
Relationship, 2, 4, 8, 18, 20, 25, 26, 28, 30, 43, 60, 61, 78–82, 84, 92, 103, 122, 131, 143, 147, 169, 170, 172, 179, 180, 184, 186

Responsibility, 2, 27, 32, 63, 102, 103, 124, 130, 131, 134, 135, 150, 152, 153, 155, 156, 159, 182
Risk, 2, 5, 9, 12, 13, 18–20, 22, 36, 52, 56, 60, 66, 72, 85–87, 89, 91, 97, 100, 106, 121–136, 151, 158, 162, 171, 178, 188, 194, 195, 199–201
Risk factor, 2, 3, 26–31, 64, 66, 80, 188

S

Screening, 5, 54, 55, 59, 76, 84, 86–90, 94, 97, 100, 109, 117, 124, 126, 129, 132, 134, 159, 160, 193, 196, 197
Supervision, 19–20, 144
Support, 2–4, 6, 7, 9, 12, 13, 17, 19, 21, 23, 25–27, 30, 32, 36, 37, 42–48, 55, 57, 62–65, 76, 81, 82, 86–88, 90–97, 101, 104–107, 111–119, 128, 131, 132, 139–142, 144, 145, 150, 158, 161, 169, 171, 179, 181, 182, 186–188, 201

T

Taboo, 9, 20
Tentative pregnancy, 61, 79–82
Termination of pregnancy, 2, 3, 5, 9, 54, 56–59, 61–62, 70, 71, 75, 77–79, 81, 84–86, 88, 90, 91, 94, 100, 107, 110, 111, 113–116, 118, 121, 122, 125, 127, 131, 134–136, 141, 142, 145, 149, 152, 155–161, 163, 167, 171–174, 181, 183, 184, 187
Trauma, 22, 23, 25–27, 186
Traumatisation, 7, 9, 13–15, 20–23, 25–28, 32, 62, 63, 69, 187, 188
Trisomy, 16, 133
Trisomy 13, 54
Trisomy 18, 17, 54, 77, 78, 161, 170
Trisomy 21, 12, 15–17, 25, 54, 56, 77, 78, 124, 126, 141

U

Ultrasonic testing, 12, 25
Ultrasound scan, 77, 90, 178, 181
Unborn child, 2, 3, 8, 20–23, 60, 71, 100, 131, 134, 150, 175